TECHNOLOGY, INNOVATION and POLICY 7

Series of the Fraunhofer Institute
for Systems and Innovation Research (ISI)

Ulrike Bross · Annamária Inzelt
Thomas Reiß

Bio-Technology Audit in Hungary

Guidelines, Implementation, Results

In Co-operation with
Rainer Bierhals, Doris Holland, Stefan Kuhlmann,
Chris Mahler-Johnstone, Silke Just and Christine Schädel

Anikó Baricz, Gyöngyi Kürthy, Zoltán Németh, Balázs Pozsony
Árpád Király, Noémi Tófalvy and Gábor Kulcsár

With 21 Figures
and 30 Tables

Springer-Verlag Berlin Heidelberg GmbH

Ulrike Bross
Dr. Thomas Reiß

Fraunhofer Institute for
Systems and Innovation Research (ISI)
Breslauer Str. 48
D-76139 Karlsruhe, Germany

Dr. Annamária Inzelt
Innovation Research Centre (IKU)
Budapest University of Economics
Múzeum u. 17
H-1088 Budapest, Hungary

ISBN 978-3-7908-1092-9

Cataloging-in-Publication Data applied for
Die Deutsche Bibliothek – CIP-Einheitsaufnahme
Bross, Ulrike:
Bio-technology audit in Hungary : guidelines, implementation,
results / Ulrike Bross; Annamaria Inzelt; Thomas Reiß.

(Technology, innovation, and policy; 7)
 ISBN 978-3-7908-1092-9   ISBN 978-3-642-52472-1 (eBook)
 DOI 10.1007/978-3-642-52472-1

Cover design: Erich Kirchner, Heidelberg
SPIN 10664806   88/2202-5 4 3 2 1 0-Printed on acid-free paper

Preface

This book offers a guideline for "Technology Audit" exercises for the transforming innovation systems of Central and Eastern European Countries (CEECs). Furthermore, the book presents the results of an exemplary application of this guideline in the field of biotechnology in Hungary. The authors - a group of innovation researchers of the Fraunhofer Institute for Systems and Innovation Research (ISI), Germany, and of the Innovation Research Centre (IKU), Hungary - provide a sound concept for the identification of technological strengths and weaknesses of the CEECs' industrially oriented research systems as a basis for the design of advanced innovation policies.

After the Organisation for Economic Co-operation and Development (OECD) had proposed a "Technology Audit" of Hungary in 1993, a pilot audit was carried out under their auspices. In parallel, the German Federal Ministry of Education, Science, Research and Technology (BMFT) put forward the idea of developing the audit concept further in order to make it applicable also in other Central and Eastern European Countries. They asked ISI to utilise the running OECD audits as a learning source and to work out a comprehensive audit approach. On the basis of the observation and analysis of the OECD audit in Hungary, under critical application of the methodological approaches used hereby, and taking the experience with the practical use of the audit results into consideration, ISI developed a concept containing several core elements: (1) differentiation between general methodological aspects applicable to "all" potential audits in Central and Eastern European countries and those which will usually have to be re-designed country by country; (2) definitions and concepts for the identification of a country's "strategic" sectors or technological competencies; (3) concepts of "competitiveness" in terms of different markets; (4) concepts for "macro" analyses taking into account a probable lack of comparable and reliable data and statistics; (5) concepts for "micro" analyses regarding the specific dynamics and the speed of the ongoing transformation processes and their consequences for the prospects of single companies or R&D labs; (6) and concepts to secure the practical "policy relevance" of the audit results.

The resulting transferable *Audit Guideline* is documented in the *first part* of this book. In order to check the reliability of the Audit Guideline it was applied to the Hungarian biotechnology infrastructure. This *case study "Biotechnology"* was carried out jointly with IKU, Budapest. It is documented in the *second part* of this book. This project serves as a good example of how Western experiences in various research areas, here especially methodological knowledge of ISI in the fields of the analysis of technological competitiveness and innovation potentials, evaluation research and international developments in biotechnology, can be fruitfully combined with specific expertise and know-how of countries in transition contributed by IKU.

Furthermore the co-operation offered the opportunity to engage in a mutual learning process between partners from Western Europe and from economies in transition on which further co-operation and exchange can build.

We owe many thanks to all participants in the various stages of this comprehensive "Audit Exercise"; they contributed lots of valuable ideas: in particular, we want to thank our former colleague Dr. Doris Holland for her engagement during the early stages of the project, the authors of the above mentioned sectoral audits, the audit initiators at the OECD, in particular Mr. Jean-Eric Aubert, the initiator of the studies presented in this book, Mr. Michael Széplabi of the German Federal Ministry for Research and Technology (BMBF, successor to BMFT), the supporters of our work at the Hungarian National Committee for Technological Development (OMFB), in particular Mr. Sandor Bottka, and last but not least, the many Hungarian experts in the field of biotechnology without whose collaboration the biotechnology audit would not have been feasible.

Frieder Meyer-Krahmer, Director

Stefan Kuhlmann, Head of Department "Technology Analysis and Innovation Strategies"

Contents

X

List of Figures

List of Tables

Summary

Technology Audit of Central and Eastern European Countries

Since 1990 the Central and Eastern European Countries (CEECs) have experienced a radical technological restructuring. Their task is to integrate into the world market. While some CEECs can rely on a good research base, the major legacy of the soviet system is outdated technology and capital stock. The instrument of technology audit aims at identifying industrial and technological strengths and weaknesses in order to support policy makers in designing an appropriate strategy for the restructuring of science and industry and technology policy.

In the review "Science, Technology and Innovation Policies Hungary" (1993a) the OECD recommended a technology audit for the country. As the Hungarian "National Committee for Technological Development" (OMFB) was highly interested in the project, the OECD carried out a pilot technology audit in co-operation with research institutes from Germany, Finland, France and Austria from April to December 1994. The OMFB selected four sectors which they considered to have potential competitive strengths: medical equipment production, packaging industry, agricultural machine production, and plastic processing industry. On behalf of the German Federal Ministry of Education, Science, Research and Technology (BMFT) the Fraunhofer Institute for Systems and Innovation Research (ISI) performed a monitoring research to analyse the audit action, especially the utilisation of the audit results in Hungary.

Within the framework of German-Hungarian Scientific Technical Co-operation representatives of both countries agreed to continue and further develop the audit activities. In 1995 and 1996 ISI performed a technology audit of biotechnology in Hungary in close cooperation with the Hungarian Innovation Research Centre (IKU). In contrast to the sectoral audits, the aim was to assess a technology field. This approach allowed for taking into account the characteristics of modern technologies which often tend to alter the links and borderlines between traditional industrial sectors. Besides its generally perceived high potential as one of the critical technologies for the 21st century, the Hungarian partners chose biotechnology because of the country's assumed strengths in this area.

The overall objective of the monitoring research carried out by FhG-ISI was to design a methodology for technology audit which could be fruitfully transferred to other transition economies. Contemporary research on technological strengths and weaknesses of economies and the practical experiences during the OECD sectoral audits served as important input to develop a transferable technology audit concept. This concept was then implemented and further elaborated through the biotechnology audit of Hungary. As a result ISI developed Guidelines for technology audit in

CEECs. Policy makers as well as possible evaluators can use them as a handbook to design a technology audit tailored to the needs of the respective country or region.

Though the notion of technology audit for CEECs now is considerably wide-spread, a broadly applicable technology audit concept for transition economies needed to be formulated. As technology audit has to identify future areas of competitiveness, its approach has to be dynamic relying on a thorough understanding of the interactive nature of the innovation process and the resulting requirements for the R&D system. It therefore has to assess not only technological standards and human resources but the interaction between the different actors of the innovation system. In advanced market economies, there exist a range of different approaches to assess the strengths and weaknesses of innovation systems and to identify innovation potentials and future technological competitive factors. Among these are approaches originating from the area of evaluation research, the analysis of regional and national innovation potentials and the assessment of a country's technological competitiveness as well as the technology audit of individual enterprises.

To transfer the methodology applied in advanced market economies to CEECs a careful adaptation is required since the conditions for technology audit in transition economies differ radically. First of all, information needs and framework conditions for technology policy in CEECs have to be identified. Especially in the early years of transition, severe problems were associated with the availability and reliability of data and the limited possibility to interpret technological or economic trends in the transformation context. To be successful in less developed innovation systems, technology audit has to develop specific indicators which mostly differ from those employed in more developed systems.

The guidelines describe the process of technology audit highlighting crucial decisions and show alternative solutions depending of the country's specific requirements. The main issues discussed relate to audit objectives, the choice of auditors, the methods of data collection and analysis, the generation of audit results and recommendations and - being of high importance - the dissemination and implementation of audit results. The methodology developed is not a rigid prescription but represents an inventory of possible audit methods which addresses as well research interests as practical questions of policy makers and evaluators.

Biotechnology Audit

Background

Besides the potential of biotechnology to provide new products, processes and services, two additional characteristics of biotechnology are mainly crucial for the assessment of the Hungarian performance: firstly, biotechnology is a universal enabling technology which can be used within different industrial sectors at different innovation stages. At present this may already be seen within the pharmaceutical industry, where the development of new drugs is almost impossible without using biotechnological methods. Secondly, modern biotechnology is considered as a basic technological precondition for the further interdisciplinary development of other critical technologies. This means that numerous important areas of technologies will be influenced by biotechnological approaches. An example of this feature is the relation between biotechnology and information technology, where a new interface between these different technologies has been developing recently: the field of bioinformatics. In general this increasing intertwining of different areas of technology will become an essential feature of the technological development at the beginning of the 21st century (Grupp 1993).

The evolution of biotechnology in Hungary is linked with particular problems whose impacts need to be taken into consideration during this analysis. Among others, the transformation of the whole economy, radical changes in the environment and markets, the collapse of the CMEA market must be mentioned. A general limitation in financial resources also creates obstacles not only for the effective promotion of biotechnology. Furthermore, considerable changes have taken place towards technology and innovation policy during the transition period.

Goals

Against this background the Technology Audit of biotechnology aims to produce the following:
- an analysis of industrial and technological strengths and weaknesses of biotechnology in Hungary in an international comparison;
- an analysis of framework conditions for biotechnology in Hungary;
- deductions for technology policy options and recommendations for the OMFB.

A mix of different methods has been applied in carrying out the Biotechnology Audit:
- evaluation of scientific literature and research studies;
- utilisation of scientific, technical and economic indicators;
- interviews with experts from research institutions, firms, governmental bodies and other institutions.

International Development of Biotechnology

A brief review of the state of biotechnology in selected countries was made to provide a baseline for the analysis of the Hungarian situation. The analysis concentrated on the United States, Japan, Germany, the United Kingdom and Israel. In order to describe the situation of biotechnology in these countries the following issues were considered:

- significance of biotechnology within the respective country,
- research institutions in biotechnology,
- commercial activities,
- framework conditions like policy measures, legal framework, financing conditions etc.

The USA are clearly the leading nation in biotechnology, as indicated by several criteria (highest number of dedicated biotech firms, broadest and strongest research base, best patent position). This judgement is especially true for health care related biotechnology. The role of the government in the past has been mainly twofold: support of basic research related to biotechnology and development of favourable framework conditions.

At least two key elements which contributed to the success of biotechnology in the USA can be identified. Firstly, there was a strong science push which could draw on a well-developed, strong research base. Secondly, the technological and commercial potential of this science push was realised rather early by the private sector. Mainly two groups are important in this context: one group are the entrepreneurs who took the risk of exploiting the commercial potentials of biotechnology. The other group are private investors who risked investing private capital into this new business. So far for both groups their engagement paid off.

It is questionable, however, whether the American situation could be transfered to other countries. This holds true especially for the second part of the success story, the venture environment. Therefore, it might be necessary to develop other strategies for transfering science into commercial success. But it is also very clear that all these exploitation strategies cannot be successful if the first factor or prerequisite, a broad and strong science base leading to a permanent science push, is not available.

Biotechnology in Japan has been driven mainly by large enterprises and trade companies. Small and medium-sized companies have played no role. The strategy of the large enterprises in the past has been mainly to acquire basic knowledge generated in the United States and thereby enter the innovation system at a rather advanced stage. This strategy has proved to be successful in many other industrial sectors. However, in biotechnology so far this approach has failed. The Japanese experience in biotechnology points towards a more general conclusion. Research-intensive technologies with a strong science link require, on a firm level, the availability of own competitive research capacities, which provide the competent interface for communicating with research sites like universities where the needed basic knowledge is generated. Strategies which try to shortcut these links are not appropriate for biotechnology.

Compared to the United States, biotechnology in Germany adopted a different course. The German biotechnology industry comprises not only start-up firms, but also many established firms which diversified into biotechnology. In addition, the business focus of industry is more diverse: besides the health care sector, environmental applications, the agricultural sector and also the industry supply sector are covered by these companies. This different industrial structure has to be considered when comparing the German situation and competitiveness with the United States. It is not very meaningful to look just at the number of dedicated biotech firms, rather the whole picture has to be taken into account.

After doing so, the situation of biotechnology in Germany can be evaluated rather positively. There is a broad and strong R&D infrastructure, personnel is highly qualified, the German and European target markets are attractive. The legal framework is now considered mostly as acceptable, public funding is available, not only from federal, but also from state and European sources. Main problems are private financing and the speed and efficiency of know-how and technology transfer. Considering the strong relation between science and commerce in biotechnology it is not only important for the future to improve the availability of private capital for biotechnology and the transfer of know-how from science to industry, but also to maintain a broad and competitive research base. In general, the whole process of innovation needs to be checked for bottlenecks which then in turn give clues for further optimisation.

Biotechnology in the United Kingdom combines elements which can also be observed in other European countries with typical features of the American situation. For example, the industrial structure in biotechnology, as in Germany, comprises one half of mostly start-up companies and the other half more established companies which have diversified into biotechnology. Political measures as in other European countries try to improve technology transfer and also increase the awareness of other industries for the potentials of biotechnology. The research base in life science in Great Britain is very strong as in the United States and also as in some other European countries. Private financing in Great Britain is considered the best developed among the European countries. This situation, however, has also led to similar problems as in the USA: the stock market, for example, is reacting very sensitively to success or failures of clinical drug developments. Another problem for the future of biotechnology in Great Britain is seen in an increasing brain drain to the United States which is also facilitated by strong similarities in the two countries.

The biotechnological industry in Israel is considered to be advanced and competitive from the scientific and technical point of view. However, many of these companies are facing problems in the transition from the R&D stage to the market. These problems are mainly due to a lack of capital and difficulties during market access. This situation could finally pose a danger that Israeli expertise in biotechnology will be recognised by foreign firms, who will then purchase it, bringing the employment, exports and profits to a grow outside Israel. On the other hand, a growing recent interest of US firms in Israeli biotechnology companies has led to the formation of strategic alliances and mobilisation of private capital for the Israeli biotech industry.

Biotechnology in Hungary

The present Biotechnology Audit by its nature was a co-operation between two institutes: IKU Innovation Research Centre and FhG ISI Fraunhofer Institute for Systems and Innovation Research. Both institutes are independent from the government and without any business interests. The research team of the present Biotechnology Audit consists of national and foreign evaluators.

The bi-national research team - German and Hungarian - is an appropriate combination. Hungarian experts are well acquainted with the situation in Hungary and since Germans are not involved in everyday discussions on Hungary they can take a bird's-eye view of the country and they have more up-to-date information on world tendencies in the field of biotechnology. Another important character of the mixed team is that members have different scientific backgrounds: biotechnology and economics. Interdisciplinarity of experts is also important. According to our experiences, scientific competence in technology and in economics are key conditions for proper audit results. Collaboration characterised all phases of the process: defining the aims of the analysis, selection of sector, developing interview guides, selection sample, fieldwork, and analysis.

Experts from policy, industry and science were brought in during the audit as interviewees and consultants. Contrary to the pilot audit, appropriate Hungarian government institutions were not involved in this follow-up pilot audit as evaluators. This made the investigation and evaluation of their role in the field of biotechnology possible. On the other hand, it was more difficult to carry out the audit and to obtain available data.

A great deal of information which is easily available in Western countries is still lacking in transition economies. Consequently, the analysis had to start at a very early information stage. The first step was to prepare a list of actors in biotechnology.

The core elements of the survey were interviews with firms and institutes and their own documentation. The investigation was conducted mainly in English. Most of the interviewees were fluent in English. Most interviews were carried out by a mixed bi-national team from IKU and FhG ISI.

Firm and institute management had open attitudes to the interviews. Participation was active in those firms that were willing to take part in the audit, similar to experiences during the pilot audit. During first contacts with firms it became clear that many firms had dropped biotechnology in the context of privatisation and re-deployment.

The interview guides developed for this audit contained open and close-ended qualitative questions and close-ended quantitative questions. The results of the interviews were recorded in detailed interview guides. On this basis very comprehensive and detailed information is available.

The main lesson of methodological discussion is that the investigated sample allowed to give an overall picture of the Hungarian biotechnology sector, even if written questionnaires have not been used and some firms refused to give interviews. We were able to cover the most important players, and contacted non-interviewed firms in some way, too. Results of the Biotechnology Audit were discussed with representatives from interviewed firms, institutes, and public bodies, from the OMFB and the BMBF during a workshop in Budapest. In general, the findings of the audit were supported by the participants of the workshop and the process of the audit itself was seen as a positive and informative experience.

The following section presents the basic features of the sample. It then highlights the main findings of the present investigation. All in all, 20 firms, 18 institutes and 8 experts were interviewed. For statistical analysis 17 institutes' and 18 firms' interviews could be used.

The detailed analysis of the audit could concentrate on two areas where biotechnology is applied: pharmaceuticals and agriculture. It is not by chance that there are only a few pharmaceutical firms in the sample and those are relatively small firms. The sample is strong in agriculture. This sector was the key targeted sector of the Large-scale Governmental Programme. This sector was able to survive regardless of turbulent changes in ownership, form of organisation, management etc. Large Hungarian pharmaceutical firms on the other hand have very limited commercialisation in biotechnology, R&D activities were diminished. Most of these firms are quitting biotechnology.

The results of the audit indicate that Hungary may expect a good position in *agro-food biotechnology* while other biotechnology-related sectors look less promising. The agro-food biotechnology sector could be a strategic one for Hungary. On the other hand it is very doubtful whether biotechnology in the pharmaceutical industry will have strategic importance in Hungary. This does not mean that this sector should be neglected. However, since it is important for success to concentrate the critical innovation mass, government policy has to choose and select target areas and target groups for its supporting measures. In this context the audit provides an additional tool for priority-setting in policy formulation.

Strengths and Weaknesses of Hungarian Biotechnology

Creation and diffusion of new technologies involves a dynamic interaction between firms and their environments. These environments, called "innovation infrastructures", consist of networks of suppliers and customers, support services, financial institutions, as well as the publicly supported science and technology infrastructure, such as universities, and public R&D programmes.

The present Technology Audit highlighted some strengths and weaknesses at both micro and macro levels. One of the most critical problems is the country's weak knowledge distribution power and coherently the inadequate utilisation of available knowledge by firms. For example, there is a wide gap between modern techniques employed in institutes and those applied in firms. The reasons for isolation and underdeveloped co-operation among potential partners of the innovation process were discussed in detail. As a consequence commercialisation of the results of earlier research and experimental development is very slow.

In principle we share the opinion of many of our respondents that *biotechnology , or more specifically certain sub-areas of biotechnology like agro-food applications, could be a strategic issue for Hungary*. However, this does not mean that everything related to biotechnology has strategic importance and all R&D costs have to be financed by public sources.

The role of the government in this context is on two different levels: the first one comprises the design of an adequate framework, especially through legislation, so that environmental conditions are favourable for research, creativity, entrepreneurial spirit, financing. Besides setting the right incentive scheme, influencing the actual environmental conditions is only possible in the long run and dependent on many other factors which partly are not within the influence sphere of the government. Furthermore, the government has responsibility in those areas where market failure occurs. In the case of biotechnology, this is true for basic research and the education system. Both constitute important inputs into the innovation process but their private provision is unrealistic. Among the success factors for biotechnology state-funded basic research has ranked very high, as shown by the US and UK as well Japanese experiences. Accordingly, government is the sole or an important player with respect to the following success factors: research base, manpower, legal framework.

Secondly, government has to intervene in the short run, to compensate for the unfavourable conditions for biotechnology. This is the government policy most often pursued internationally. It is a logical consequence to explore and appropriate the research results. Success factors which need support measures are especially know-how transfer and interdisciplinary networking and financing. In this context, international experience with different policy instruments, e.g. practices of direct and indirect funding in Germany, could be used when developing respective policy measures.

In general, many problems hampering the development of biotechnology originate in the global economic situation like budget cuts, high interest rates due to high inflation, underdeveloped purchasing power, not fully effective company restructuring and others. The unstable Hungarian market and changeable economic conditions are not favourable for strategy formulation and strategic analyses are usually not well founded. In the following policy implications and recommendations based on the results of this audit will be summarised, taking account of these difficulties. These will touch on three levels: general issues; framework conditions for biotechnology; research institutions, business and innovation networks.

General Conclusions

- In general terms, the privatisation process has to be finished, re-deployment of organisations by the state has to be settled soon. Imperfect legal regulation should be improved and rules made clear.

- Concerning the future of biotechnology firms, we have to emphasise that the financing system of R&D and innovation are key issues for the Hungarian market economy model. Two large public funds (OMFB and OTKA) can hardly fulfil their tasks because the shortage of money is a great obstacle to efficient working. There are no clear priorities if we investigate the declared aims from the distributing side of state (funds) financial sources. Industrial research carried out and financed by businesses is on a small scale. In-house expenditure on R&D including biotechnology is very limited. Spending on in-house research is important even if firms do not aim at producing new scientific results. According to international experiences, if a firm does not have any R&D capability it is hardly able to adopt (acquire) new technologies. Even follower companies need some R&D capacity if they wish to adapt effectively. Summing up firms' opinions we do not have too many illusions regarding business-financed projects. Some of them are waiting for new results and would be glad to commercialise right away, but they think government has to cover R&D expenditure at institutes. Business has to learn that R&D output from research institutes is not available without costs. If firms would like to remain players on the biotechnology stage they should consider R&D partners more seriously. If they need new R&D results to improve their competitiveness, they have to invest in them.

- One of the current strengths of the Hungarian biotechnology sector is the good quality of human resources. But at the same time there are not enough well-educated specialists. Both government and business have the duty to create more possibilities (or at least to maintain the existing ones) for *education and training of young specialists*. If no one invests in education and post-graduate courses any more, the quantity of qualified people will become one of the greatest impeding factors of further development.

Conclusions Related to Framework Conditions for Biotechnology

- It has advantages and also disadvantages that a specific *biotechnology law* has not yet been enacted. While such a legislation might create obstacles for performing certain biotechnology activities which are under legal constraints in other countries, a missing special biotechnology framework on the other hand leads to an uncertain and changeable situation with negative impact on planning security in firms. In addition, if Hungary wants to become an international player in biotechnology, it will be necessary to adopt international legal standards. This applies, for example, to the enactment of the long-discussed biotechnology law, and also to the international convention on animals. In addition, revision of slow and costly approval procedures in the pharmaceutical and in some cases in the food industry is needed.

- Until now there has been no public debate about biotechnology in Hungary and there is no lobby against biotechnology, which can be considered as a strength and opportunity of the technology. The country can avoid the emergence of a negative public attitude towards biotechnology if public knowledge is increased in time so that the public can decide, on a well-informed basis, about e.g. accepting biotechnology related products.

- For both policy-makers and industry an *up-to-date data bank* would be helpful as an information base in order to develop and implement policy measures and firm strategies more efficiently. This follow-up audit invested a lot to identify biotechnology organisations. Its results offer a good starting point to develop an up-to-date register and data bank which could complement the register on agriculture-related institutes which has been prepared recently by the Ministry of Agriculture.

- Insufficient *demand for biotechnology, know-how, products and processes* is a burdensome factor for institutes. Relations between the science and industry sectors are weak but neither are co-operations within the industry sector strong. If the lack of co-operation persists, achieving a competitive position in biotechnology will be impossible given highly flexible and fast acting international players. Furthermore, the lack of co-operation may lead to the loss of potential partners in the future. Government can encourage university/industry linkages, offer support to establish technology transfer organisations and create new possibilities *to join European programmes*.

Conclusions Related to Research Institutes, Business and Innovation Networks

- *Benefit of intellectual assets* may become a source for further development and to this end, attention should be paid to patents. The attitude of inventors towards patent applications has to be changed. This is not only a task for government, but for R&D organisations, too. Institutes need to strengthen their patent activities and marketing, and make a *selection in their research portfolio,* taking into account criteria related to the scientific value of projects (mainly basic research organisations) and to value and demand (mainly applied research and experimental development organisations). Institutes are waiting for firms as partners to finance their research projects and experimental development. However, their "supply" has to meet business demand.

- As the SWOT analysis highlighted, firms and institutes have to *invest more in marketing* and try to *develop links* with each other and with the international community, too. They have to broaden their information sources, concentrating more on business-type information. Institutes have to pay more attention to business demand when offering R&D services. Agricultural institutes are usually ready to meet firms' demands. They are more client-oriented than other research organisations, but they also have to face the problem of a lack of demand. Some basic research organisations e.g. university departments, declared they are ready to continue applied research or experimental development on a contract basis to earn money to develop their laboratories and to receive normal wages.

- In many cases R&D organisations need a *new style of management.* Management of scientific organisations is a problem, not only in transition economies, but all over the world. At the end of the 20th century a brilliant scientist is not automatically the appropriate leader for a research organisation. The problem goes deeper in transition economies than in advanced market economies because market-oriented practices such as fund raising were out of the question until the late years of socialism.

This audit helps to clarify the nature of the "Hungarian genius" as it is manifested in biotechnology science and technology and its potential contributions to the economic and social development of the nation, which now finds itself following major political change, involved in intense global economic competition. This strategic effort could also be useful to prospective foreign investors and partners, because it would reduce the often prohibitive "search cost" for them. In addition, this Biotechnology Audit may contribute to a development which had been initiated by the OECD pilot Technology Audit namely that technology audit will become a part of Hungarian policy thinking.

This audit also provided a lot of methodological results and perceptions. It may improve the efficiency of the dissemination of technology audit knowledge if domestic institutes can participate in every phase of the work. Co-operation between two institutes was a significant element of the follow-up audit and at the end of the project we could evaluate it as beneficial. The core part of the method is relevant for investigation of other sectors. It is an important task to make technology audits which are conducted in different sectors at the same time as comparable as possible. The general conceptual elements can serve as a Technology Audit Guideline for Central and East European countries. It may disseminate the technology audit experiences in other Central and Eastern European countries, either in training seminars or with audit actions.

I. Technology Audit Guidelines for Central and Eastern European Countries

1. Introduction

The transition from a centrally planned to an open market economy affects not only all aspects of the political and economic systems of the Central Eastern European Countries (CEECs), but has to be carried out moreover under highly unfavourable economic conditions. Since the early 1990s, governments in CEECs have achieved considerable successes in terms of political and macroeconomic stabilisation. Experiences have proved that there is no single strategy appropriate to handling the complexity and uncertainty of transition.

The objective of the present research project is to support governments by the development of Guidelines for Technology Audits in CEECs. Technology audits support the formulation of strategic policy for long-term economic growth. The majority of economies in CEECs are far from consolidation, therefore, the need for a long-term strategy for economic development becomes evident after the completion of basic institutional restructuring. Science and technology play a key role for industrial restructuring and long-term economic growth, innovation and technology policy should be an integral part of the overall economic policy aiming at the mobilisation of existing technological strengths. Even in an early state of transition, political neglect can cause permanent damage of innovation potentials.

However, the process of policy formulation and implementation encounters more difficulties in CEECs than in advanced market economies. The responsibilities of policy-makers need clarifying: while the formerly pursued top-down approach of policy has proved to be ineffective, in the field of science and technology a pure bottom-up approach appears not to be viable either. Furthermore, to have the intended impact, policy measures have to be founded on a thorough evaluation of the situation and require full insight into the functioning of the system.

In 1993, the OECD expressed the need for a technology strategy for Hungary, recommending a technology audit for the country in order to evaluate the technological strengths and weaknesses (OECD 1993a). The present research project is complementary to the OECD action for Hungary but also has a more general scope, since it aims at providing an analytical instrument for other CEECs. To develop guidelines for technology audit, two different lines of research will be pursued: on the one side, the range of possible methodologies will be explored to constitute an inventory of different approaches as exhaustive as possible. On the other side, careful research has been devoted to the transferability of the methodology to CEECs. To name only a few, possible difficulties with data availability arise due to lack of or incompatible, official statistics or practical constraints for empirical field studies.

Chapter 2 presents an overview of recent political and economic developments in CEECs and describes the overall characteristics of the system of innovation. Since the notion of technology audit - as introduced by the OECD - is relatively new, the concept and objectives of technology audit will be clarified in Chapter 3. Alternative methodologies for technology audit can be derived from a survey of international contemporary research on technological strengths and weaknesses of advanced market economies, but also of CEECs. The experiences made during two pilot actions in Hungary, i.e. an OECD Audit in four industrial sectors in 1994 and a Biotechnology Audit in 1996, are presented in Chapter 4. They provide valuable insights into the organisation of the audit process as well as in the transferability of research methods to CEECs.

Both the literature survey and the empirical evidence constitute the basis for the concluding Chapter 5 which presents the Guideline for Technology Audit in CEECs. The orientation is towards readers less interested in scientific evaluation of alternative methodological approaches but more in concrete advice. For further comments on the finally selected best practice for technology audit, the preceding chapters provide additional information. Researchers involved in the audit may use the final chapter as a handbook. Besides the presentation of different methods for the collection and analysis of data, considerable emphasis is placed on the organisation of the audit process.

2. Situation in Central and Eastern European Countries

2.1 Political Transition and Economic Performance

In the transition from a centrally planned to an open market economy all CEECs have achieved major progress. The political transformation seems an irreversible process. After a period of drastic decline of production output and GDP in the early 1990s, economic stabilisation and positive growth rates of the economy have been achieved. The future role of CEECs in international trade relations will depend strongly on their ability to foster and complete the process of economic restructuring and to create industries which can flexibly adopt to market demands. Long-term economic perspectives depend on the consolidation of existing technological assets and the exploration of innovation potentials. As a group of CEECs have formally applied for membership of the European Union, their future economic performance can well affect also the present members of the Community.

Geographically, Central and Eastern European Countries comprise all countries located to the west of Russia. They represent a heterogeneous group of countries in terms of initial conditions, strategy for transformation and present performance. Therefore, the present chapter does not attempt to give a complete overview of recent trends in CEECs, but points out some key characteristics of the economic situation and concentrates on main patterns of science and technology in CEECs. Countries belonging to the Community of Independent States (CIS) are not considered explicitly because of their diversity from CEECs. Nevertheless, some of the patterns described hold also true for CIS and the results of the research project can be fruitfully explored for this group of countries.

For all CEECs, macroeconomic data suggests that the bottom of economic development has been in the years 1991 or 1992 and since 1994 positive growth rates of GDP have been achieved (cf. figure 2.1-1). After periods of three or four digit inflation rates in some countries, monetary stability has strongly improved recently. According to inflation rates in 1995 three groups of countries can be identified: a rate around 10 % p.a. has been achieved by Albania (8.0%), Croatia (4.1%), Czech Republic (9.1%), Slovak Republic (9.9%), Slovenia (12.6%). The second group ranks around 30% p.a., including Hungary (28.2%), Poland (27.8%), Romania (32.3%), Estonia (29.0%), Latvia (25.0%) and Lithuania (35.0%). The third group with Bulgaria (62.0%) and Macedonia (50.0%) has still high inflation rates (World Bank 1996, p. 174). In spite of this positive macroeconomic outlook, changes in real wages compared to the previous year have not been positive in all CEECs: in the Czech Republic and in Slovenia real wage changes have remained positive since 1992 and 1993, in Poland, the Slovac Republic, Romania and Lithuania since 1994 (BMWi 1996).

Figure 2.1-1:   GDP Growth Rate (Percent)

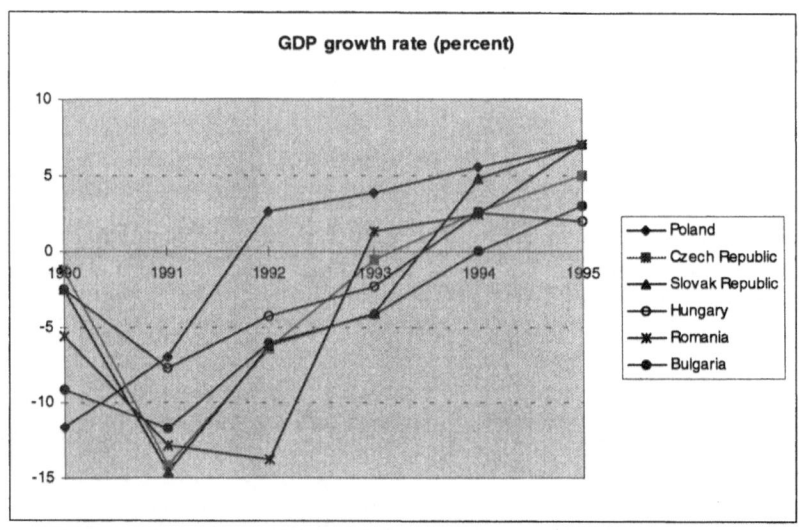

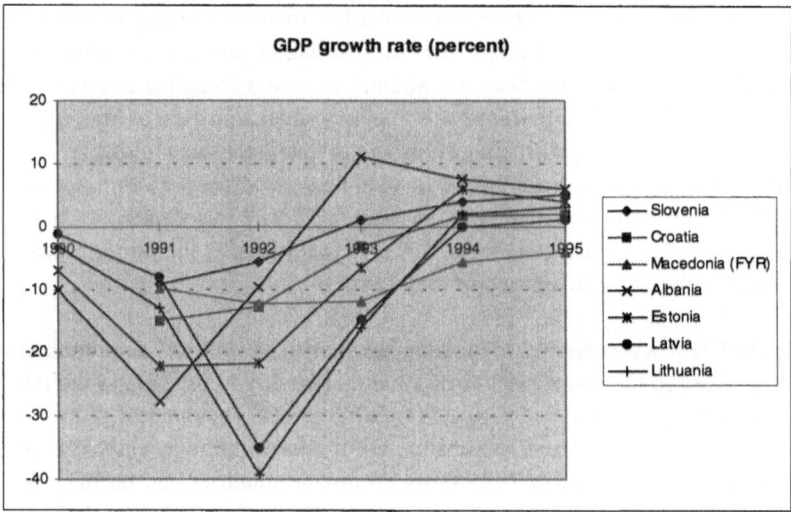

(Computed on the basis of World Bank 1996, p. 173)

The majority of CEECs has adopted new constitutions while the approval of civil and commercial law is under way. The initial conditions for transformation varied among CEECs: the foundation of new states represented both a chance and a challenge to create new democratic institutions. Other CEECs had to transform the public bodies of their former socialistic states but at the same time could rely on an existing system.

Basic institutional reforms have been completed which include the liberalisation of prices, factor markets and foreign trade. The restructuring of the banking systems has been started in most CEECs but often encounters considerable difficulties because of persisting state involvement, huge loan losses and dry capital markets. On the contrary, the privatisation of production facilities has made considerable steps forward after some delays in the past years. The percentage of private sector of GDP is illustrated by the figure below:

Figure 2.1-2:   Share of Private Sector of GDP in 1995

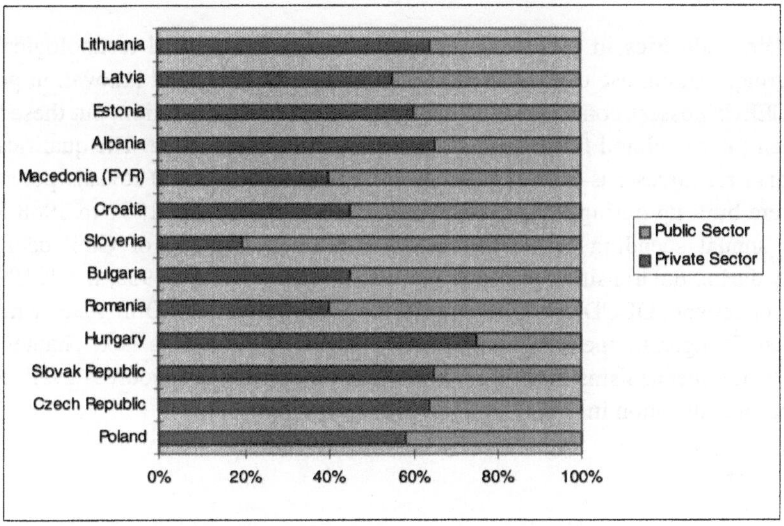

(Croatia, Macedonia, Latvia and Lithuania: data from 1994, Slovenia: data from 1992, Estonia: estimated)

(Computed on the basis of BMWi 1996)

The increase in private sector activities indicates the considerable progress in economic restructuring which has been achieved. But a brief reference to statistical problems associated with the data presented illustrates persisting critical issues: national accounting in some CEECs includes formally privatised enterprises, which still belong to the state, into the definition of private sector activities. Furthermore, the informal sector has increased between 1989 and 1994 with an estimated annual rate of 18 to 22% in CEECs (World Bank 1996, p. 27). Consequently, a considerable share of productive activities does not appear in national statistics.

Macroeconomic stabilisation and institutional reform are necessary conditions for a permanent economic development in CEECs. After the collapse of production and trade relations in the former Council for Mutual Economic Assistance (CMEA), CEECs need to develop a competitive position in international trade. But the initial conditions are far from promising. Industrial structures and levels of technological development which are ill-adopted to market requirements are the legacy of decades of central planning in the CMEA context. To give examples, production tends to be inefficiently resource-intensive and quality standards are often low. The CEECs suffer from a technological gap since the socialist system hampered flexibility of industries and economic growth. This is further aggravated by the fact that industries in advanced market economies are characterised by the emergence of global competition with technological development as a key factor.

Therefore, industries in CEECs require both organisational and technological restructuring, making use of existing technological capabilities and innovation potentials. CEECs possess considerable strengths in science and education, but these have not been fully explored for improving technological performance. The qualification of human resources has been high, also as expenditure in R&D reveals potentials that were built up during the socialist time. For example, from 1981 to 1988 Hungarian annual spending on R&D reached at average 2.44 % of GDP declining sharply during the transition period to 0.99 % in 1993 (OECD 1993a, Inzelt 1995b). As a comparison, OECD countries spent 1.95 % of GPD on R&D in 1993, which is less than Hungarian spending in the 1980s (OECD 1996). The next chapter will focus on the mechanisms that led to the underemployment of resources available in science and education in CEECs.

2.2 Modernisation of Science and Industry

Innovation can be defined as the development of technologically new or improved products or techniques and their commercialisation in the market or implementation within production (Meyer-Krahmer 1987, p. 317). In CEECs, process innovations harness technological restructuring through introducing new resource-saving techniques into production. Product innovations responding to market needs represent a possibility for industries to compete in international markets and survive in home markets against foreign companies.

Innovations result from the application of scientific knowledge in industry or the combinations of existing know-how. Market success requires demand-orientation and the willingness to incur risks. In order to employ the innovation potentials of CEECs in an economically successful way, the functioning of the national systems of innovation needs considerable improvement.

The organisation of science had a comparable structure in all CEECs: main actors were the national Academy of Sciences, universities and branch institutes. The main player in basic research was the Academy of Sciences which consisted of a large number of affiliated research institutes. Universities were primarily engaged in teaching while involved in research only to a very limited extent. Branch institutes whose activities were in applied research formerly belonged to industrial complexes and were therefore under the control of the respective branch ministry. Applied research was also performed in industrial research units within companies. Furthermore, a considerable proportion of research was devoted to military purposes with its own apparatus.

In the organisation of research, a top-down structure was predominant. Very little co-operation took place within the science sector and between the science and industry sector. Therefore, the potential in the research sector which existed in the majority of CEECs was not applied in industrial production. Since the beginning of the transition, major restructuring has taken place, not only in industry, but also in the research sector, while countries have followed very different strategies. In most countries, the formerly existing potentials have suffered severely from the lack of financial resources, the dissolution of industrial research units and the associated brain drain. Consequently, many research institutes were not able to keep up their standards while on the other hand the pace in science and technology in advanced market economies has increased.

The problems behind the reluctance to co-operate are not only lack of resources but also a systematic behavioural pattern. Research institutes did not perceive their role as problem-solvers for industrial production. As a result, only very little research was oriented towards the needs of industry. Faced with the need for restructuring the research system, associated with changes in resource allocation, the segmentation and reluctance to co-operate among research institutes was even intensified during the past years. As international experience shows, the ability to network, share scarce resources and compute new information beneficially is an important property of a successfully operating innovation system.

A closer look at modern trends in technological change clarifies the requirements which innovation systems have to meet in order to strengthen industrial competitiveness. First, the technology intensity is growing in high-tech sectors but second, also mid-tech industries are increasingly affected by science-based technological developments. Third, a major source of innovation is the fusion of heterogeneous technologies leading to the emergence of new disciplines and growing interprenetration of technologies (Kuhlmann/Reger 1996). These developments represent considerable challenges for industry, especially for small and medium-sized enterprises (SMEs): while on the one side they function as an important element of the technological dynamics of a country's industries because of their flexibility and vicinity to clients, on the other side they have to cope with size disadvantages in terms of scarce financial and human resources.

Therefore, firms have to be able to explore the possibilities of acquisition of knowledge - which crucially depends on the internal R&D management - and need to rely on a qualitatively excellent research system. The easy access to the science base and interdisciplinary networking is important. Ideally, the innovation system provides a long-term oriented network of existing institutions being competent to offer technological but also management support to industry, especially in selected strategic areas. This has to be complemented by a system of education and vocational training to provide enterprises with highly skilled human resources (Kuhlmann/Reger 1996).

The consolidation of existing scientific and technological capabilities in CEECs depends on a consistent strategy for technology and innovation which has to be an integral part of the overall economic policy for growth. After the assessment of the specific conditions in the respective country, such a political strategy serves as basis for the formulation of concrete policy measures. Technology and innovation policy aims at the generation, accumulation and diffusion of technological know-how and their economic application. Therefore, it is directed towards the different producers and users of technological knowledge, which are in science and industry.

Technology and innovation policy has to comprise instruments such as funding of public research institutes, financial incentives for industrial research activities and the building or extension of the innovation support infrastructure, as well as support for young businesses. The design of such a technology strategy has to rely on the specific conditions of the respective country. In the following chapters, technology audit will be presented as an instrument to support technology and innovation policy in CEECs.

3.   Principles of Technology Audit

3.1   Technology Audit as an Instrument of Innovation Policy

In this chapter a brief summary of the main aims and underlying principles of technology audit with special reference to countries in transition will be given. On this basis, technology audit will be defined. Then international experience of related types of analysis and their respective methodologies will be reviewed. Finally contemporary research on issues related to technology and innovation in CEECs and the methodologies which have been employed in a transition context will be presented.

In the review "Science, Technology and Innovation Policies Hungary" (1993a) the OECD recommended a technology audit of the country, in order to identify technology areas with specific strategic advantages. The OECD aims to provide a foundation for the elaboration of technology and industrial strategy. Emphasis is on the important contribution of science and technology to the transformation process.

The definition of technology audit as formulated by the OECD for Hungary will be the starting point for further investigation into the conceptual and methodological aspects of technology audit in CEECs. According to the OECD Secretariat the process of technology audit has the following objectives (OECD 1995a):
- Identify competitive strengths and weaknesses
- Discover policy options at both government and firm level
- Facilitate foreign investment.

The audit focuses on sectors of the economy in which technological standards as well as human resources are evaluated. Its scope is extended towards potential technological strengths, as it examines not only technologies already in use in industry, but also laboratory prototypes and early development (OECD 1993a and 1995a).

Although the notion of technology audit for CEECs is now considerably widespread,[1] a broadly applicable technology audit concept for CEECs has yet to be formulated. It would be necessary to develop a specification of the objectives and information requirements of such an audit. In particular, the proposed level of analysis, possible criteria for identifying strengths and weaknesses in a given sector and scope of policy options would need to be clarified (Holland 1995).

---

[1] In the editorial of the July/August issue of Business Central Europe, a review published by The Economist Group, the authors suggested a technological audit of the Czech industrial base which would "identify the country's capital need and its skill shortage" (p. 5).

Technology audit provides data required for the formulation of innovation and technology policies. The audit concept has to take into account specific characteristics of innovation systems in the individual CEECs. Innovation and technology policy is directed towards the development and exploitation of the technological strengths of a country. Therefore, such policy is crucial during the transformation of an industrial sector as well as in helping to promote an environment conducive to long-term competitiveness. National systems of innovation have not yet overcome the legacy of the soviet system in terms of organisational structure and mentality. This creates obstacles which prevent the flexible functioning of the system necessary to support industrial requirements (cf. Chapter 2).

For data collection and for the formulation of policy options, the proposed technology audit has to be designed in such a way that it reflects an understanding of both the innovation process and the functioning of the innovation system. The nature of innovation is necessarily interactive. Feedback links connect different stages of the innovation process. Kline and Rosenberg described the chain-linked model of innovation: in addition to a central chain-of-innovation which "begins with a design and continues through development and production to marketing" (Kline/Rosenberg 1986: p. 289), a second path consisting of a series of feedback links is introduced. There is a direct connection between perceived market needs and users' requirements to potentials for improvement of product and service performance in the next round of design. In addition, there is a relationship of frequent exchange of ideas between the research sector, which contributes to the stock of knowledge, and the development sector. This model describes innovation within firms or between different actors in the innovation process.

The previous description of the innovation process illustrates how the transfer of scientific results in the commercial innovation of a marketable product or process may impose high demands on all actors involved. The extent of these demands are influenced by the effectiveness of the functioning of the national system of innovation, a concept introduced by Freeman and Lundvall. Freeman defines the national system of innovation as "the network of institutions in the public and private sector whose activities and interactions initiate, import, modify and diffuse new technologies" (1987: p. 1). Lundvall identifies the following elements of the national system of innovation (1992: p. 13): internal organisation of firms, inter-firm relationships, the role of the public sector, the institutional set-up of the financial sector, R&D intensity and R&D organisation.

The previous discussion of the conceptual background leads to the following working definition of technology audit for CEECs:

Technology Audit is the assessment of the technological and organisational strengths and weaknesses of all relevant actors in the innovation system of a country or a region in CEECs. It also relates to their exchange relationships and to the framework conditions under which they operate. The analysis has to take into account the interactive nature of the innovation process. The dynamic perspective allows the identification of technological potentials in CEECs accounting for their specific situation of restructuring and new orientation of the industries. In practice, technology audit has to focus on pre-selected sectors or segments of the economies in CEECs.

A conceptual approach towards technology and innovation guides the design of the technology audit in terms of level of analysis, data collection, indicators selected; these and other methodological questions will be discussed in the following chapters.

3.2   Contemporary Research

3.2.1   Introduction

The idea of technology audit for CEECs as developed in the previous section was only recently introduced. It relates to various established approaches of policy evaluation, analysis of the innovation potentials of regions, the technological competitiveness of a country and the strengths and weaknesses of individual companies which have a well-established tradition in advanced market economies[2]. In this context, the aims and possible methodologies of technology audit are not new. To fulfil its aim of providing a comprehensive analysis of strengths and weaknesses and the generation of policy recommendations, technology audit should acknowledge these approaches and can build on contemporary evaluation practices. The first section of chapter 3.2 will discuss these four sources for technology audit. The discussion will help categorise the elements of technology audit. In the following, German and international research relating to the analysis of technological strengths and weaknesses in general will be considered. Then selected studies of the technological situation in CEECs will be surveyed.

---

[2]   These categories are neither meant to be exhaustive nor clear cut since it has to be acknowledged that there are many overlapping types of analysis.

3.2.2 Evaluation Research

"Evaluation research is the systematic application of social research procedures in assessing the conceptualisation and design, implementation, and utility of social intervention programs." (Rossi/Freeman 1985: p. 19). The systematic assessment of US governmental social programmes in the 1930s can be viewed as a landmark of early evaluation research (Rossi/Freeman 1985: p. 21). The origins of evaluation of science and technology include research into the control of public spending, assessments of science and technology within the scientific community, social impact research and attempts to control research and development expenditure (Kuhlmann/Holland 1995a: p. 13).

There can be identified two main lines of development of the evaluation of science and technology: at an earlier state, evaluation functioned primarily as an internal control mechanism for the scientific community (Georghiou 1995: p. 9). Procedures employed at this stage were peer-review and the measurement of research performance of individual researchers and groups (e.g. using bibliometric methods) as a basis for the allocation of funds to research institutions. Complementing this science-internal concept, the second line of development consists of science-external evaluation studies which asses the impact of science and technology policy programmes (Kuhlmann 1995: p. 26). Evaluation research is most commonly conducted on specific government programmes, to assess the needs of the target groups of government programmes and the future potential of new technologies, but is also applied to the knowledge infrastructure. Especially the latter form of evaluation providing a basis for institutional reforms and the restructuring of organisational missions and making scientific and technological institutions more responsive to their clients represents a new kind of evaluation (Kuhlmann 1995).

The evaluation of technology programmes is initiated by policy-makers to prove the achievement of scientific, technological, economic or societal targets. Programme evaluation is an instrument not only to validate the existence of such actions but to assist the fine-tuning of R&D programmes. Evaluation units are mostly independent, but may also be bodies within the government administration. The formal procedures for implementation of the results and their actual impact may differ according to the country in which the evaluation was undertaken and according to the type of evaluation chosen (i.e. ex ante, ex post and monitoring evaluation). As objectives and their associated methods are highly heterogeneous, there is no common evaluation practice. For instance, a variety of possible approaches towards institutionalisation and evaluation practice is persisting in the countries of the European Union (cf. Research Evaluation 1995). Kuhlmann/Holland (1995) present the range of evaluation projects carried out for the Federal Ministry of Education, Science, Research and Technology in Germany. A good overview over possible evaluation methods and their application by given in Bozeman/Melkers (1993).

Case study: Evaluation of the knowledge infrastructure

This novel concept was initiated by the government of a federal state: The Ministry for Economic Affairs of Baden-Württemberg supported the development of a generally applicable qualitative and in-depth evaluation procedure for the 35 industrially oriented R&D institutes which are co-funded by the ministry.

The approach is based on two elements: The first element consists of nine 'success factors' which would help the managers of R&D institutes to run their lab effectively: strategic orientation; technology management; links with industry; links with science; communicative competence; organisation and management; human resources; techno-scientific equipment; financing.

As a second element, five 'performance criteria' would help the funding government bodies to assess the actual fulfilment of the tasks of the institutes: coherence of strategic business area planning and need of the relevant industrial sector for technological support in the future; techno-scientific competence; success in finding economically viable solutions; income base; human resources and techno-scientific equipment. These criteria will be used in concerted evaluation processes with independent external evaluators and representatives of the government on the one hand and the institute under evaluation on the other.

This evaluation concept is based on the idea that the effectiveness and the future performance potential of an industrially oriented R&D lab cannot be assessed simply by the number of scientific publications or by the size the institute's income from contracts with industrial corporations. Rather, it aims at a qualitative evaluation of an institute's performance potential and of the pre-conditions for future success.

Source: Kuhlmann 1995, Kuhlmann/Holland 1995b

3.2.3   Innovation Potential of a Region

The findings of certain branches of innovation research which have emphasised the crucial role of the spatial dimension for innovation and firms' development have now parallels in industrial and innovation policy of regional and national governments as well as of the European Union. The geographical vicinity of firms plays an important role, not only in terms of transportation costs, but moreover a common culture and learning relationships between regional agents take place in spatial structures. The term of the innovative milieu was developed by the GREMI school (Aydalot 1986, Camagni 1991). The European Union aims at the strengthening of less developed regions within the community through programmes like "Regional

Innovation Infrastructures and Technology Transfer Strategies" (RITTS) and the "Regional Technology Plan" (RTP).

The practices and tools employed in regional planning are manifold. Nevertheless, the analysis of regional innovation potential is a core issue for regional development. Main objectives of analysis relate to the demand side and the supply side of the regional innovation system, especially the identification of firms' needs and the evaluation of the innovation infrastructure (Muller/Gundrum/Koschatzky 1995: p. 1). As such analyses are oriented towards potentials and have a strong policy focus, the assessment of existing strengths and weaknesses have lower priority. In spite of the specific question, there is considerable closeness to ex ante evaluation research as described above. Similar to technology audit for CEECs, the analysis of regional innovation potentials prescribes not one single approach but combines different methodologies (Nauwelaers/Reid 1995).

3.2.4 Technological Competitiveness of a Country

The evaluation of the performance of industries or the whole technological system of a country (or a region or federal state) may serve as an indication of the need for political action. Therefore, the results are addressed to governments, politicians, and industrial associations. Most commonly, technometrical indicators such as patent activities and export data are employed referring to past achievements as well indicating future opportunities.

The assessment of the innovation potential and the associated competitiveness of a country's industries has to be based on the measurement of innovation activities and the identification of the determinant factors for innovation. These were captured by Schwitalla (1992) who performed a horizontal econometric analysis of enterprises' innovation behaviour in Germany. She relied on innovation indicators such as R&D expenditure, R&D personnel, patent applications, public funding and investment exploring published annual reports of enterprises and patent data banks. As competitiveness or eventually the persistence of a "technological gap" in the international context is a matter of economic structure, the analysis has to consider a country's performance in high-technology sectors (Gehrke/Grupp 1994). These are associated with especially high future potentials. The development of an up-to-date list of high-technology products allows for the evaluation of the specialisation and diversification patterns of R&D, patent activities, share of high-technology of the economy and foreign trade patterns of a country in international comparison.

A further methodological refinement was introduced by Münt (1995) who took into account new developments of international exchange such as export of services, FDI, firm-internal flows and intra-industry trade. To explain long-term structural change, especially the relation between accumulation and technological gap, the author chose a longitudinal analysis. As an example, the strategic study of Schmoch/Grupp/Laube (1996) on the technological performance of Swiss industries represents the application of the research on innovation activities and long-term competitiveness highlighted above. The analysis elaborates future technological trends and the challenges for industry which arise from these developments.[3] This kind of analysis can benefit from results of technology foresight as a different but related policy instrument.

An alternative analytical grid for the analysis of technological strengths of countries has been put forward by Porter in his "Competitive Advantage of Nations" (1990). The author distinguishes four determinants of the competitiveness of a country: (1) factor endowment, (2) strategy, structure and rivalry of firms, (3) demand conditions and (4) related and supporting industries constituting the so-called "diamond model". Porter's research is based on case studies of ten industries in ten different countries.

Case study: Development strategies for the Swiss industry

The aim of the study was to provide concrete and ready to implement strategic recommendations for the development of new product clusters. Recent data on international technological developments and the performance of the Swiss industry were analysed in four steps.

Step one designed a framework of technological development which was the basis for elaborating potential future technology targets in the form of product visions (*step two*). This selection was generated through the analysis of international technical trends considering as well competitive developments, price movements and market demand involving discussions with experts. *Step three* assessed the current technological performance of the Swiss industry identifying their future strengths and weaknesses on the basis of a technological and economic profile. This led to the recommendation of strategies and technology policy measures in *step 4*.

On the ground that transdisciplinarity will be a dominant feature of future technological developments the study focussed on the ability of the Swiss industry to appropriate the synergies associated with the merging of technologies. This ability depends on the matching of the industrial profile with future opportunities and threats. The analysis of the technological profile was performed on the industry level but also at cluster and firm level and employed indicators such as R&D budget, R&D personnel, patents and foreign trade patterns. World market shares were compared with patents on the most important world markets, furthermore the structure of patent holders were analysed.

---

3   In this context see also Grupp (1993) and Faust et al. (1995).

Criteria for the evaluation of the future competitiveness of the Swiss industry were the level of innovativeness and the technological direction. As one of the main results, a dominant share of innovation activities of Swiss industry were in technological areas with low science linkage which can represent a misfit to future challenges.

Reference: Schmoch/Grupp/Laube 1996

3.2.5   Technology Audit of Enterprises

Technology audit originally is a concept referring to individual firms (Saage/Hemer 1981, Saage et al. 1992, Muller/Gundrum/Koschatzky 1995). An in-depth analysis of the technological capabilities, procedures and needs of individual firms is carried out (Muller/Gundrum/Koschatzky 1995: p. 31). The aim is to develop, diversify and increase the productivity and effectiveness of enterprises (Schneider 1992: p. 1), for example through the formulation and evaluation of diversification options, systematic market research, innovations in the manufacturing system or in co-operation relationships. Technological and business experts visit the enterprise several times involving management as well as functional personnel. The objectives of the audit focus on the resource inputs and functioning of the firm's internal organisation while the intensity of the analysis may depend on the specific assignment. The approach is strongly implementation-oriented as the results of the intensive audit lead to recommendations responding to the firm's needs. Therefore, audit results are addressed to the internal stakeholders.

Technology audit requires high technological and business expertise and can benefit from developments in strategic management theory to conceptualise the audit results. One major device at the stage of defining strategy is the so-called SWOT analysis, i.e. the identification of strengths, weaknesses, opportunities and threats. This concept classifies internal vs. external factors and actual vs. potential determinants of the competitive position of a firm. Internal strengths and weaknesses restrict the possibilities for action of the enterprise, while future environmental factors form a point of reference to assess whether the actual capabilities match the arising challenges. The SWOT tool can be employed to prepare an in-depth analysis as well as to condense and convey the findings of the thorough audit (European Commission 1995: p. 261, Bartol/Martin 1994: p. 172).

Also, technology audit can be used as a tool for professional investment decisions, for instance by venture capitalists. This device can be extended to identify commercial potential of research projects and research portfolios of institutions in general (DesForges 1992: p. 6/7).

All these approaches to technology audit of firms have an important point in common: the evaluation of strengths and weaknesses of individual enterprises or entities always takes into account relevant market trends and the development of production technologies, or compares the actor's performance to his or her industry.

Technology audits of firms can be used as a policy instrument, as in Ireland where the National Technology Audit Programme was launched in 1989 to improve the performance of manufacturing industry (European Commission 1995: p. 9).

3.2.6   Summary and Methodological Implications

By comparison to the approaches presented, technology audit for CEECs is not revolutionary new. It has the ambitious aim to identify the specific rationale of the functioning of the innovation system and its deficiencies in CEECs, and therefore embraces various elements of the types of analysis presented. From their different characteristics, we can derive a set of categories which serve to further describe the methodology of technology audit.[4]

Methodological Questions of Technology Audit
(1) Target group of audit results
(2) Subjects of investigation
(3) Level of aggregation of (2)
(4) Auditors
(5) Type of indicators employed
(6) Data collection method(s)
(7) Data analysis method(s)
(8) Reference point(s) for comparison and evaluation
(9) Time horizon
(10) Frequency of analysis
(11) Procedures of dissemination and implementation of audit results

Experiences in advanced market economies constitute an inventory of possible instruments and indicators for technology audit. The range is presented in the following table.

---

[4] Meyer-Krahmer has proposed a typology of evaluation research in the field of innovation policy distinguishing the following categories: type of evaluation, issues, data and methods, frame conditions (Meyer-Krahmer 1991: p. 23).

Table 3.2-1:   Summary of Methods to Analyse Technological Strengths

	Different Approaches
Target group of results	Policy-makers, regional planners
	Firms, research institutes, investors
	Public in general
Subjects of investigation	Actual and potential strengths, weaknesses
	Tangible, intangible assets
	Needs analysis
	Impact analysis
	Actual and potential diffusion of technologies
	Basic assumptions of policies
	Framework conditions
Aggregation level	Countries
	Industries, individual enterprises
	Research infrastructure: universities, institutes, related institutions
	Government programmes
Auditors	Independent institutions, units within the government administration
	Management and technical experts, methodological expertise
Type of indicators	Subjective, objective data
	Input, output, process related
	Retrospective, prospective
	Quantitative, qualitative
	Macro, micro
	Established, emerging
	Representative, explorative
Data collection method	Direct, indirect
	Exploitation of existing data (statistics, studies)
	Postal questionnaires, interviews, case studies
	Expert workshops
	Panel, data banks
	Representative samples and control group analysis
Data analysis method	Qualitative analysis
	Cost-benefit-analysis
	Econometric models, technometric and bibliometric methods
	Peer review
Reference point for comparison	Country-wide and international comparison
	Technological trends, product visions, foresight techniques
	Before-after-analysis
Time horizon	Horizontal, longitudinal analysis
Frequency of analysis	Once, periodical, permanent
Implementation of results	Concrete recommendations, policy scenarios
	Validation of results
	Seminar and workshop
	Publication
	Consulting, follow-up visit

In the following section and chapter the existing studies on technological development in CEECs and the pilot audits in Hungary will point to conclusion, which methodology can be recommended for CEECs. Options in terms of different methodologies will be presented in the form of a Guideline in the concluding Chapter 5.

3.3   Research on Technological Strengths in CEECs

3.3.1   Introduction

This chapter presents a selection of studies relating to industrial and technological strengths in CEECs. A guideline for technology audit in CEECs cannot be the mere transfer of a methodology which is successfully employed in advanced market economies. The suggested methodological options can build on already existing research on the national systems of innovation in CEECs.[5] Previous research can point out important open questions and provide a general understanding of the specialities of the respective country. The survey of literature on CEECs constitutes an inventory of data sources, methods, indicators for the assessment of technological strengths and weaknesses. It reveals the possibilities and difficulties in employing the methodologies presented in the previous chapters in the context of transition.

3.3.2   Selection of Literature

After the fall of the iron curtain, research topics have focused on political change and macroeconomic issues such as inflation, employment and stabilisation. Since the early 1990s microeconomic research related to the transformation and restructuring of the economy, especially privatisation, investment and emergence of new actors and industrial and innovation policy has been carried out. From this rich body of literature documents are selected according to their relevance for technology and innovation, i.e. which investigate the individual actors of the innovation system and their relationships as well as the system as a whole. A second selection criteria is to present a heterogeneous range of analyses. At the same time it has to be acknowledged that any kind of representative choice per se seems to be impossible. The documents taken into account range from data computed and released by official bodies (e.g. OECD, National Statistical Offices) to working papers. They serve as examples for different objectives and methodologies. To capture the specialities of the studies, three broad working categories are distinguished.

Categorisation of Documents
1. Aggregate analysis
2. Enterprises, their internal organisation and inter-firm relations
3. Knowledge infrastructure

---

5   In the following, literature on the transformation of political and economic systems includes as well studies on the former German Democratic Republic wherever appropriate.

3.3.3   Aggregate Analysis

The aim of many aggregate analyses is to give an overview of the economic and political development in CEECs in order to better understand and appreciate the starting point and achieved stage in the transformation process.

Recently, information on the political and economic development and on the issues relevant for R&D has significantly improved in almost all CEECs, thanks to the international action of the OECD, the World Bank, the EU and national reporting activities (World Bank 1996, OECD 1993a, BMWi 1996, EU 1996, Weidenfeld 1995). These documents provide quantitative data on mostly macroeconomic indicators, allowing a benchmark of the CEECs among each other or internationally. Mostly, the documents analyse elements of the legal framework, i.e. property rights, business law, trade regulation, foreign exchange, tax regulations, actual administration and its enforcement and the institutional set-up of the financial sector. Partly, they are computed on the basis of reported data of CEECs. Widespread practice is to involve CEEC experts and to organise workshops to collect the data and to validate the research findings. Case studies are employed to much lesser extent (see e.g. OECD 1993a).

Research on the transformation process of the economies in CEECs employs economic theory (e.g. Carlberg 1994, Glismann/Horn/Stanovnik 1995, Hauer et al. 1993, Hickel/Priewe 1994, Indruch 1994, Klodt/Paqué 1993, König/Steiner 1994, Siebert 1993), but also makes use of political science and sociology (Pradetto 1994, Schimank 1995), especially theories on transformation of systems (Fleissner 1994, Gehrig/Welfe 1993, Szántó 1995). Technological aspects of transformation have been analysed e.g. by Bentley (1992) for the former GDR and by Schneider (1994) for the Soviet Union. As a rare exception Hernisniemi/Hyvärinen (1995) investigate technological potentials of the Baltic states with reference to innovation theory.

Research studies have been qualitative and quantitative. Among the publications there is a huge share of anthologies and workshop proceedings. Quantitative analysis employs macroeconomic indicators such as economic growth, industrial production, Foreign Direct Investment (FDI), specialisation patterns in terms of exports and production. R&D statistics report input and output indicators, among them bibliometric information: Gross Domestic Expenditure on R&D (GERD), R&D labour force and R&D scientists and engineers, education indicators, patent activities, especially the number of patents granted in the US, citations, and publications. The aggregation level of analysis is mostly very high, only few studies employ disaggregated data. For example, ifo (1991) analyses production patterns of CSFR and Poland in deep disaggregation aiming at identifying technologies which represent investment opportunities. The methods employed range from technometric and bibliometric analysis to macroeconomic simulation models (Braun/Schubert 1997, Gehring/Welfe 1993, Guerrieri 1994, Roeger 1994). The range of countries which are used as points of reference for evaluating the performance of CEEC economies is remarkably wide: EU, US, Japan, the Tiger Countries, developing Latin American countries and other CEECs.

The analysis has to capture on the one hand real adjustments associated with the profound restructuring of the industrial production and the economic sector. On the other hand statistical problems, especially difficulties with unharmonised industry classifications, have to be overcome. Therefore, aggregate quantitative analysis has certain drawbacks as indicators can only rely on very recent trends with limited capacity to predict the future technological potentials of CEECs (Hutschenreiter 1994). Overlapping effects cannot be isolated and process indicators which might hint at future developments are extremely difficult to assess in such an aggregate analysis. In the literature some attempts are being made to increase the data reliability through employing patents granted in the US or considering trade balance with market economies (e.g. Guerrieri 1994). Validation of results with CEECs experts is sought frequently (e.g. Weidenfeld 1995).

To conclude, the information base has been considerably improved. This is especially true for some countries like Hungary, the Czech Republic and Russia, to a much lesser extent for other CEECs however. Qualitative and quantitative studies have provided an overview of the economic development of CEECs or single aspects of the transformation. Concerning the technological strengths and weaknesses, insights in the political and social logic of the R&D sector in the CMEA context and its transformation have been provided. But a gap remains between the explanatory and analytical power of qualitative analysis and the foundation of arguments through quantitative indicators. The assessment of technological strengths and their future potential is in its infancy, as well as recommendations or concepts for industrial and innovation policies.

While research at an aggregate level is an important input into technology audit of a country, the in-depth understanding of restructuring and innovation potentials in the business sector requires per se a micro-oriented analysis.

3.3.4   Research on Enterprises, their Internal Organisation and Inter-firm Relations

Research interest as well as research opportunities related to the transformation of the business sector through privatisation, foreign capital and emergence of new firms has increased significantly (see Brezinski/Fritsch 1996). This research comprises studies on the private business sector, especially small and medium-sized enterprises, of a whole country but also research on specific aspects relating to enterprise development. With different aggregation levels the possible perspectives on the individual agents of the system of innovation vary. The analysis of the enterprise sector is often based on empirical fieldwork.

Examples of the analysis of the SME sector in Poland, the Baltic States and Bulgaria are Smallbone/Venesaar/Piasecki (1996) and Stoyanovska/Krastenova (1996). Among others, studies focus on specific questions such as investment-relevant issues, especially the role of FDI (Borsos 1994). A considerable share of research is also dedicated to the internal restructuring and market orientation of companies (e.g. Dolgopiatova 1996). A more disaggregated analysis of the management of restructuring in Russian industries has been carried out e.g. by Lipsitz (1996) and Shama (1995), for Bulgaria see Shapira/Paskaleva (1994). Several studies analyse selected industries, Auvinen (1994) carried out an analysis on the Russian software and hardware industry, Shaw (1996) on the Russian aerospace industry. These studies relate to economics, but also make use of organisational business theory.

In all these studies, the identification and assessment of technological strengths is under-represented. For instance, the role of SMEs in transition economies is highlighted as important for short-term job creation rather than seen as a source for future economic competitiveness. When considered at all, issues relevant for R&D are mostly investigated as an integral part in the overall context of enterprise restructuring. Only some analyses have the development of R&D in the business sector as their main focus. Hilbert (1994) analyses the R&D strategies of companies in the transformation crisis of Eastern Germany. Her focus is on the reorganisation of R&D organisation, employing both quantitative and qualitative indicators. Kaiser/Tamm (1992) present a case study on R&D and technology management in CEECs. Albach/Witt (1993) analyse the network of suppliers and customers of former state-owned enterprises in the GDR, emphasising the importance of access to western information networks. Joint research of FhG ISI and University of Freiberg on the restructuring of R&D organisation and the innovation potential in Saxonia (former GDR) analyses among other subjects the innovation behaviour of firms (Fritsch/Bröskamp/Schwirten 1996). Several studies on Hungary were carried out by the Innovation Research Centre (IKU) at the Economic University of Budapest (e.g. Inzelt 1995a 1996c). An overview of future perspectives for CEECs' industries with reference to the important role of technology and innovation is provided by Widmaier/Portratz (1996).

As regards the methodological findings of the surveyed documents, qualitative studies are more predominant than quantitative work; only recently has quantitative research been increasing. The methods of analysis employed range from analytical presentation to econometric analysis and simulations models. Frequently used data sources are existing studies. Direct data collection methods such as interviews, postal questionnaires, even longitudinal analyses have been used as well. Sample size varies from rather small groups to almost representative numbers. Those documents which are written out of the perspective of business theory strongly employ case studies to account for specialities. Only a smaller part of these studies are theory-driven, referring to evolutionary models or innovation theory (e.g. Inzelt 1996a, Keren 1996).

To conclude, empirical research has contributed to a better understanding of the determinants of the restructuring of enterprises and - sometimes less explicitly - the R&D process within enterprises and whole industries. On this basis, technological strengths and weaknesses can be evaluated. But the research does not yet allow a comprehensive perspective of the respective industries and national economies. This would be the necessary precondition for well-founded policy recommendations. Concerning methodological aspects, reliable information can be gathered through direct data collection methods which generate promising results. These analyses contribute substantially to an inventory of methods which can be employed in a technology audit in CEECs.

3.3.5   Research on the Knowledge Infrastructure

The technological strengths of the knowledge infrastructure and the role of the public sector in technological development in CEECs has not yet been widely investigated. Studies mostly analyse out of a more general perspective the legacy of the soviet-style science and technology system (e.g. Dyker 1994) or concentrate on the political economy of the research system (Mayntz/Schimank/Weingart 1995). One example for the technology oriented analysis is a contribution assessing a science park in Russia (Batstone/Westhead 1996) or an investigation of the Hungarian research infrastructure (Tournemine/Muller 1996). In several countries, first attempts at evaluating the knowledge infrastructure have been made, for example in Slovenia, Hungary, Romania and Ukraine (Balazs 1994, Inzelt 1996b, Isakova 1996, Segal Quince Wicksteed 1995). So far, there seems to be no systematic literature on the knowledge infrastructure. Approaches in Hungary in the framework of the OECD and Biotechnology Audit will be reported on later in Chapter 4.

3.3.6   Summary and Methodological Implications

The following table summarise the methods of analysis employed in the documents which have been surveyed.

Table 3.3-1:   Summary of Methods to Analyse Economic and Technological
                 Position of CEECs

	Different Approaches
Target group of results	Research orientation
	Capital investors
	International public, OECD and EU member states
Subjects of investigation	Transformation of the economy, esp. private sector
	Needs/constraints analysis
	R&D, technological and innovation potentials
	Framework conditions (macroeconomic and institutional)
Aggregation level	Countries
	Industries, individual enterprises
	Research infrastructure
Auditors/authors	Scientists
	International organisations/donors
	Involvement of CEEC experts
Type of indicators	Subjective, objective data
	Domestic, foreign
	Input, output data
	Retroperspective
	Quantitative, qualitative
	Macro, micro
Data collection method	Direct, indirect
	Exploitation of existing data (statistics, studies)
	Postal questionnaires, interviews, case studies
	Expert workshops
	Varying sample size
Data analysis method	Qualitative analysis
	Econometric models, technometric and bibliometric analysis
Reference point for comparison	Other CEECs
	EU, Japan, Tiger countries, developing Latin American Countries
Time horizon	Horizontal, longitudinal analysis

Research on CEEC economies and industries employs almost the full range of
methods of data collection and analysis representing the state of the art of technology audit as summarised in Chapter 3.2. While technology audit for CEECs can use
the research results on the respective countries as well as the methodological experiences, major work remains to be done. On the one hand, pitfalls of the methodology in terms of availability and reliability of data exist. These aspects are more
technical in nature. On the other hand, the methodology to assess future technological perspectives is rather underdeveloped. These open questions will be briefly addressed in the following.

To point out some of the problems persisting in the analysis of industry in CEECs:
the comprehensive identification of the actors in the innovation system of a country
or a region is extremely difficult. This stems from the highly dynamic process of
change and different legal procedures in CEECs. As an example, the computation of
a representative sample of small and medium-sized enterprises in manufacturing has

to take into account various aspects: national statistical bureaux have data of diverging compatibility with the research questions. While some countries require the registration of small and medium-sized enterprises, these figures only roughly estimate the activities and the actual number of enterprises. Some individuals may find it only attractive to register because of associated tax holidays. In some countries a national clearing house (e.g. in Croatia) reports on all financial transactions between enterprises. Still, the firms and their activities in the grey or black market can hardly be estimated. On the side of public infrastructure, name changes and altered responsibilities have to be found out.

When comparing the analyses on CEECs surveyed above with the international state of the art of technology audit (cf. Chapter 3.2), the need for further research becomes clear. Given considerable heterogeneity among authors of the studies, their analytical approach differs widely. Most research is eclectic in nature. The comprehensive analysis of the economy or selected industries in a country are very resource-intensive, in terms of expertise and financial means. Because of this heavy resource requirement, the more comprehensive studies are either carried out by international organisations such as the OECD or the World Bank, or are sponsored by the EU within the Phare framework or other donors (e.g. Deutsche Forschungsgemeinschaft), especially when involving empirical fieldwork.

Moreover, only few analyses of technological strengths in CEECs have been carried out. The indicators employed for analysis in CEECs are less varied than in advanced market economies. Especially process indicators and information revealing future potentials are mostly not employed. Reference points for evaluating the performance of industries in the transformation context have to be formulated more precisely. Furthermore, the role of public infrastructure and government activities are little explored. Research is mostly scientifically oriented, without orientation towards possible policy measures to be implemented.

Technological strengths are surveyed either on highly aggregate level of analysis or on an individual level; these insights have not been put together in a consistent mapping so far. The aim of technology audit is to rely on detailed analysis and generate representative and valid results for a country or more precise, selected industries of a country.

4.   Pilot Technology Audits in Hungary

4.1   OECD Technology Audit in Hungary

4.1.1   Background and Concept of the OECD Audit[6]

This chapter describes the background of the OECD Technology Audit which was carried out in Hungary during the second half of 1994 (June-December). The focus will be on the process leading to the decision itself and on the selection of the subjects of the audit.

After the OECD recommended a technology audit for Hungary in 1993, the Hungarian government expressed their interest and the need for such an action. At a first meeting in February 1994, which was organised by ISI in Karlsruhe, the OECD agreed to provide the conceptual and organisational frame for four sectoral audits (cf. table 4.1-1).

Table 4.1-1:   Time Schedule of the OECD Audit in Hungary

Date	Content
1st Meeting *8 Feb. 1994,* *Karlsruhe*	**Discussion of possibilities for Technology Audit in Hungary:** Participants: Federal Ministry of Education, Science, Research and Technology (BMFT, Germany), National Committee for Technological Development (OMFB, Hungary), OECD, ISI, representatives of institutes and public administration
2nd Meeting *14/15 Apr. 1994,* *Budapest*	**Draft of concept and time schedule for Technology Audit** Assignment of sectoral studies: (1) Agricultural machine production (2) Medical equipment production (3) Packaging industry (4) Plastic processing industry
3rd Meeting *12/13 Sept. 1994,* *Budapest*	**Discussion of methodological approach and first results of sectoral audits, as well as of concept for monitoring research**
4th Meeting *30/31 Jan. 1995,* *Budapest*	**Discussion of final sectoral audits**

---

6   Chapter 4.1 relies partly on the Interim Report: Evaluation und Transfer eines "Technology Audit"-Verfahrens für MOEL - Strategisches Projekt im Rahmen der Transformationshilfe für Ungarn (1994 - 1996), Zwischenbericht, Ergebnisse der Begleituntersuchung des OECD-Audits in Ungarn, Dr. Doris Holland, Fraunhofer-Institut für Systemtechnik und Innovationsforschung, Karlsruhe März 1995.

After the first meeting, the OMFB selected four sectors in which Hungary presumably possessed assumingly technological strength. The sectoral audits were assigned to the following five research teams: Agricultural machine production: Institute for Advanced Studies (Vienna), Medical equipment production I: Fraunhofer-Management-Gesellschaft (Munich), Medical equipment production II: Institute Bertin (Paris), Packaging industry: Fraunhofer-Management-Gesellschaft (Munich), Plastic processing industry: VTT (Helsinki). In order to co-ordinate the five different audit reports, the OECD presented the "Common Methodological Principles and Structure for Sectoral Reports of the Technological Audit Hungary" (see Annex I) at the second audit meeting in April 1994.

BMFT initiated the present research project on transferable Guidelines for Technology Audit in early 1994. It was perceived as a complementary project to the sectoral audits in Hungary. The aim is to draw general lessons for technology audit in CEECs from the individual and broadly differing studies which were part of the audit in Hungary.

During the third meeting held in Budapest in September 1994 the methodological approaches of the audit teams and first results of their investigations were discussed. At the same time ISI presented the concept for the methodological analysis which was supported by OECD and the sectoral teams.

The discussion of the different approaches to audit and the results of the sectoral audits during the fourth meeting in January 1995 led to the even stronger acknowledgement of the need to design a transferable methodology for technology audit. Such an elaborated methodology can be of great value for other CEECs which intend to undertake a similar action. As a result of this final conference, the OECD Secretariate drafted a "Methodology of Technology Audit" (see Annex I) which will be referred to throughout the next chapter.

The experiences of the Hungarian audits is one important input to the development of guidelines for technology audit in CEECs. The present research builds on interviews with the audit teams and the analysis of the audit reports, focusing on the organisation of the OECD Audit as a whole and the different approaches employed by the audit teams. Moreover, the discussions during the audit conferences and the methodological recommendations of OECD play an important role.

4.1.2 Experiences of the OECD Audit

4.1.2.1 Introduction

The analysis of the OECD Audit in Hungary and the assessment of the Biotechnology Audit, which is presented in chapter 4.2, follows the analytical grid presented below. It is based on both the recommended "Methodology for Technology Audit" of the OECD and the literature survey presented in Chapter 3.

(1) Audit objectives: target group of audit results

(2) Subjects of investigation and their aggregation level

(3) Selection of auditors

(4) Data collection

(5) Data analysis

(6) Generation of results and recommendations

(7) Dissemination and implementation of audit results.

The focus of the analysis will not only be on individual elements of audit methodology but also on the organisation and process requirements.

Figure 4.1-1:   Process of Technology Audit

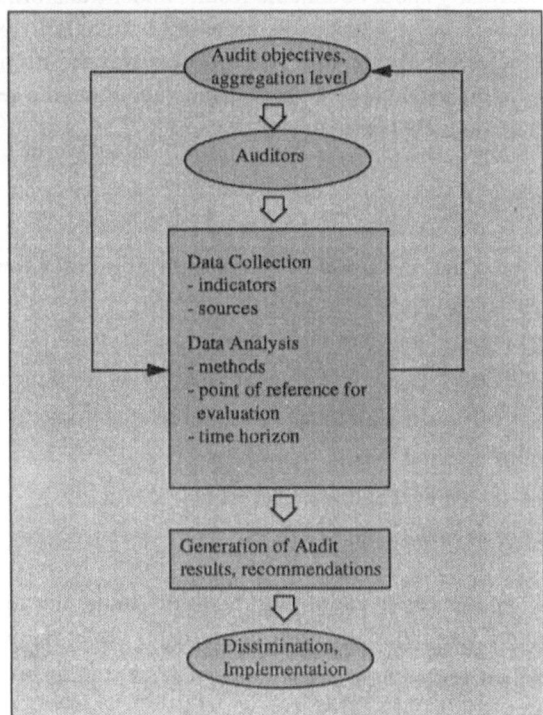

In the following, the actual procedures and experiences during the OECD Technology Audit are briefly presented, as well as the experiences and possible suggestions for improvements summarised.

4.1.2.2   Audit Objectives

The OECD formulated three main aims of technology audit (cf. Chapter 3): *first*, to identify competitive strengths and weaknesses, *second*, to develop policy options for the government and at the level of enterprises and *third*, to facilitate foreign investments.

These aims served as starting points for the sectoral studies. For a concrete audit project, the main issues and information requirements need further concretisation and adaptation to the specialities of the respective country. This applies especially to the following points:

- Criteria to select the audit subject(s)
- Definition of aggregation level of subjects investigated
- Target group of recommendations and required concreteness.

Hungarian parties, i.e. the OMFB and the Ministry for Trade and Industry (IKM), selected as subjects of the audit four potentially competitive industrial sectors. Given the high uncertainty about obtainable results under transformation conditions, the selection of different audit teams made it possible not only to explore a range of different methodologies but also to allow for different ways to formulate the leading research questions in the technology audit. This included that the sectoral teams had to choose the aggregation level of analysis.

4.1.2.3   Selection of Auditors

The OECD considered that the selection and participation of actors in the technology audit should take account of the following aspects:

- Open internationally to ensure fair competition
- Involvement of appropriate government institutions from the start
- Involvement of firms, research institutes and related institutions
- Involvement of national experts
- Involvement of technical and management expertise
- High quality of experts is the key factor.

According to the experiences made during the pilot audits, the selection of the research teams required particular attention. As a major issue, the team should be as independent as possible and should not be influenced by any own interests, particularly business interests.

It was recommended that domestic experts should be part of the international research team: in the role of independent evaluators they should be directly involved during the conception, the fieldwork and the formulation of recommendations. It was considered important to involve experts from policy, industry and science. However, the experience of several sectoral teams showed that a considerable effort was necessary to identify competent, willing and objective partners.

During the Technology Audit in Hungary, the co-ordination between OECD and OMFB proved to be very beneficial and generated constructive impulses for the field research.

4.1.2.4 Data Collection

Concerning data collection methods the OECD Methodology concept emphasised the following aspects:
  – Core elements: interviews and site visits by several experts
  – Use of local language
  – Use of international standards in data collection
  – Explicit criteria and definitions (sectors, segments, competitiveness)
  – Selection of firms should include some poor performers.

The sectoral studies carried out in Hungary were based primarily on well-established research instruments (cf. table 4.1-2): written questionnaires and interviews. Expert discussions with representatives of various institutions relevant for enterprises and research institutes took place to a limited extent. Data bank inquiries only played a role in one sector. Technometrical methods for technology monitoring and the analysis of international competition (e.g. patent analyses, export analyses as a function of technology intensity of the product) were not used.

Table 4.1-2:   Analysis and Evaluation Methods in the Sectoral Studies of the "Technology Audit" in Hungary

Analysis and evaluation method	Sectoral Audits				
	A	B	C	D	E
Written questionnaire					
Enterprises	●	●	●	●	●
Research institutions	○	•	•	○	•
Interviews					
Enterprises	●	●	●	●	●
Research institutions	•	•	•	•	•
Expert discussions (with other institutions)	○	•	○	•	●
Data bank searches	○	○	○	●	•
International comparisons	•	•	•	•	•
Use of method ● intensive (10-20 actors) • limited (5-9 actors) ○ none					

Written questionnaires proved to be a useful instrument of technology audit. The response rate was relatively high taking into account the number of firms which had gone out of business. According to the sectoral teams, the questionnaires were filled in as required. This picture had only to be revised to some extent, as on-the-spot visits revealed that some of the self-estimated responses were more positive than the actual situation warranted. Furthermore, it was recommended to issue questionnaires also in the language of the country and to accompany with a cover note from the national technology policy administration.

The organisation and conduct of **interviews** required strong support from authorities of the audit country. Wherever co-operation and commitment were evident during the sectoral studies, the interviews with firms and institutes ran particularly smoothly.

The selection of the sample of institutes and especially of enterprises, appeared to be problematic: although the Ministry provided lists of addresses and supported the initiation of contacts, relevant partners were not easily to identify, e.g. due to bankruptcies. This affected the composition of the sample negatively: it was no longer guaranteed that a mixed subset was analysed which included economically stronger and weaker enterprises. Partly, interview partners were found according to the snowball principle: visited enterprises pointed out address and contact person of other enterprises. During the discussion of some sector studies, this procedure led to the question whether a representative picture of the relevant actors and the relationships between actors was achieved by the audit.

The **combination of written questionnaires and interviews** was considered to be a highly effective approach. The method supported the selection of interview partners and also allowed adjustment for possible distortions in answering behaviour, e.g. too positive written assessments.

In order to clarify possible misunderstandings it was suggested as a possible improvement for future studies that enterprises should have the opportunity to give feedback on the research team's evaluations. Hungarian experts were usually involved in the surveys. Some sectoral studies also invited experts from other countries. Out of the perspective of enterprises and institutes which were subject to the audit, any bias caused by possible self-interest of foreign research teams or experts was a highly delicate issue. In order to avoid this problem, independent audit experts from the respective country could be nominated who would have to be involved in the selection of outsiders.

To analyse the enterprises' and institutes' environment, **interviews with experts** in Hungarian authorities and other institutions (e.g. sector-specific ministries, embassies) proved essential and were extremely helpful.

Data bank inquiries were only possible to a limited extent. This was true especially for data banks outside the respective country. Nevertheless, in some fields up-to-date data banks exist in CEECs which can be used in a very specific context, such as the ORKI data bank in Hungary in the field of medical technology. It lists all firms which have had their equipment certified for quality control, together with special data up to 1994.

In Hungary, visiting **fairs** proved very effective, as they provided an opportunity to conduct informal and open discussions with many firms.

Comparative **international analyses** were carried out by the sectoral teams, but only to a very limited extent.

Statistics have been used even to a lesser extent than under the hindering circumstances possible. This was mainly because of the time budget. Probably for the same reason, there were no systematic sectoral **literature** studies. However, particular importance should be paid to this information source in CEECs, in order to compensate for the deficits still existing in official statistics.

To conclude, since data bases did not meet the needs of the audit, gaining access to relevant data was extremely difficult. The identification, assessment and preparation of all possible data sources was regarded as a necessary prerequisite for the field-work study and therefore should be available even in advance.

The most important data sources for all sectoral studies were the interviews with firms and firms' own documentation. Most of the sectoral teams met with very open attitudes on the part of firm management to the interviewers in this respect. Only in one study did several firms refuse on-the-spot interviews. In the case of active participation of Hungarian experts in the interviews, this had strong impact on the success of the empirical research at the firm level.

4.1.2.5   Data Analysis

The sectoral studies employed particularly firm-specific indicators and data to describe the strengths and weaknesses of the sectors.

Table 4.1-3:   Data/Indicators in the Sectoral Studies of the "Technology Audit" in Hungary

Data/Indicators	Sectoral Audits				
	A	B	C	D	E
Microeconomic Data					
quantitative	●	●	•	●	•
qualitative	●	●	●	●	•
Macroeconomic Data					
quantitative	•	●	○	•	○
qualitative	○	○	•	●	●
Use of data/indicator: ● extensive • limited ○ none					

At the enterprise level, primarily quantitative data were investigated: number of employees (total, proportion of R&D personnel), turnover, production volume (including export/import) data on R&D, share of external R&D, profits and patents or licenses, share of product and process innovations, share of foreign capital.

Because of the frequent deficits and rapidly changing data, the emphasis of the sectoral studies was on qualitative description of the firms. Analysis of firms' strategies as they relate to the terms of reference of the audit also played a prominent role.

The qualitative indicators used at a microeconomic level in the sectoral studies included: range of products and processes (main products: mass products or small series), assessment of the products and processes regarding their position in the life cycle, structure of suppliers, company aims associated with products (market strategy), user structure for products and processes, competition (strategy, competitors, state of competitiveness), main technological emphasis, status of technology (price, quality, service, image, ergonomics, design), patenting and licensing activities, property relations, management of the firm, marketing, and financing.

Although all sectors focussed on microeconomic analyses, the indicators on which the data - and the corresponding results of the survey - were based were not always explicitly mentioned.

At the macroeconomic level, less data were gathered. Quantitative macroeconomic data, which were analysed in some sectoral studies, related primarily to production volume, employees, productivity, R&D expenditure and trade balance of the sector or branch. In addition, data for related sectors and branches were taken into account, e.g. to highlight the situation of users. In general, statistics with explanatory power were difficult to obtain. Those sectoral studies which analysed political, technological and economic context factors mainly employed qualitative presentations.

Besides overall statistical problems, it was very difficult to carry out economic studies given the short time available for the audit. Most of the sectoral teams pointed out the time restriction. They felt that preceding research on the history, on policies and the social and technological environment as well as analyses of the international state of the art of the technologies employed in the respective sectors and of market developments could have contributed fruitfully to the interpretation of the microeconomic research results in their context.

4.1.2.6   Generation of Results and Recommendations

According to the OECD Draft Methodology, the following requirements should be met. They refer to the generation of results and recommendations and to their dissemination and implementation (cf. Chapter 4.1.2.7).

- Equal time for evaluation and data collection
- Submit draft for comments from firms
- Iterative process with government
- Discuss effect of selection and non-answers
- Name actors in recommendations
- Follow-up actions should be embodied in process.

The time budget and the resources necessary depends on the intended size and scope of the audit. In the pilot audit the interview phase was considered extremely intensive. If different audit teams analyse sectors which are interlinked, the exchange of interim reports should be institutionalised as well as a clear division of labour defined in advance. In the pilot audit, this was successfully the case in the medical equipment sector which was studied by two different teams.

The pilot audit in Hungary showed that close contact between the actors, i.e. between survey teams and government representatives was necessary. Especially the elaboration of policy recommendations and identification of actions and possible actors required close co-operation with stakeholders from policy, science and industry.

4.1.2.7 Dissemination and Implementation of Results of the Audit

The OED distinguished three aspects concerning the information collected during the audit:

- Open information should be shared
- Confidentiality should be decided by originators
- Reports in two parts: open and closed (e.g. for firms).

Most studies in the pilot audit adhered to the "common methodological principles" of the OECD (see Annex I), but with varying degrees of depth. The sector-specific main recommendations derived could be used as an appropriate basis for follow-up activities.

The stock of information collected during the audit provided a solid basis for further analytical and advisory activities. Their relevance was twofold: on the level of policy formulation, they could be exploited together by the sectoral teams, policymakers and experts from the respective sectors. On the level of firms, the overall results and recommendations could be complemented by more concrete and action-oriented recommendations. As a result, most sectoral teams had at their disposal an inventory of unpublished, firm-specific material which would form a basis for specific advice in co-operation with Hungarian experts, if required. Possible future follow-up activities strongly depended on the priorities set by the national government.

The presentation of the sectoral studies at the OECD meetings with Hungarian experts provided a forum for Hungarian actors and helped to overcome communication barriers. As Hungarian technology and innovation policy was very much on the defensive since the general election in 1994, the OECD action helped to mobilise political actors. The strategy of OMFB was oriented towards strengthening innovation activities in Hungary and it attempted to overcome the fragmented and vertical structure of the Hungarian research and industry landscape. In this context, the OMFB regarded the technology audit as an instrument to support the transition process in Hungary. Independent research teams could hence act as mediators in the policy dialogue.

4.2 Biotechnology Audit Hungary

4.2.1 Background and Concept of the Biotechnology Audit

In order to continue and further develop the technology audit action started by the OECD, German and Hungarian government representatives agreed to carry out a "Biotechnology Audit" in Hungary within the framework of "Scientific Technical Co-operation" (WTZ) between the two countries in March and April 1995.[7] The aim of this follow-up action was to further develop the methodological lessons of the OECD sector studies for technology audit which could be transferred to other CEECs. The Biotechnology Audit pursued a **different approach from the OECD** action: the focus was on a **specific technology** field, instead of on industrial sectors. This allowed to take into account the characteristics of modern technologies which tend to alter the borderlines between traditional sectors. This is especially true for biotechnology as a **key technology**. From this perspective, the biotechnology audit was a pilot action.

The audit was carried out jointly by ISI and the Institute for Innovation Research (IKU) in Budapest. The first project meeting was held in May 1995 in Budapest.

4.2.2 Experiences of Biotechnology Audit

4.2.2.1 Introduction

The methodological issues of the "Biotechnology Audit Hungary" will be described in detail in Part 2 of the overall project: "Evaluation and Transfer of the Technology Audit Concept to Central and Eastern European Countries".[8] The following presentation follows the structure of technology audit computed on the basis of the literature survey and the recommendations of the OECD (cf. chapter 4.1.2.1).

---

7 The Technology Audit in the field of Biotechnology was agreed upon at the 6th conference of the "Gemischte deutsch-ungarische WTZ-Kommission" (Joint German Hungarian Commission for Scientific and Technical Co-operation) in Aachen (Germany) in March 1995 and subsequent consultations with OMFB.

8 The following subchapters rely to some extent on documentation of A. Inzelt (IKU).

4.2.2.2 Audit Objectives

The objectives of the audit can be further specified in terms of the target group to which the results of the audit are addressed, the subjects of investigation and their level of aggregation. The **target group** of the results of the Biotechnology Audit in Hungary were primarily the OMFB and other policy-makers who were relevant for research and business activities in the respective technology. At the same time, the results and recommendations were directed towards the actors involved in the audit.

Criteria for selecting biotechnology were its economic importance for Hungary, especially in agriculture, the pharmaceutical and food industries. Consequently, biotechnology played an important role in Hungarian technology and innovation policy and the Hungarian authorities have explicitly chosen this field for the audit. One of the underlying assumptions, which was already put forward during the first audit meeting with OECD in February 1994 (Kuhlmann 1994), was the suboptimal organisation of the actors in science and industry. This pointed towards the following **subjects of investigation**: research institutes and enterprises actively involved in biotechnological activities as well as the relationships between these actors. In terms of **level of aggregation** the empirical investigation was carried out at the level of the individual actor, i.e. firms or institutes, aiming at a consistent and representative picture of the technology in the country. Therefore, analysis on the aggregate level complemented the research on the micro level.

4.2.2.3 Selection of Auditors

The OECD Audit in 1994 was carried out by specialised institutions of OECD member countries. Hungarian experts and institutes were invited to contribute in the sectoral studies, but no Hungarian institutes were directly involved as auditors. As this would have facilitated the transfer of international best practice to Hungary, this solution was regretted on behalf of Hungarian parties.

The Biotechnology Audit was carried out jointly by ISI and IKU. Both institutes are independent research institutes without any business interest, which satisfied one very important requirement for the selection of auditors. The co-operation was close throughout the whole project, i.e. definition of aims of the project, elaboration of the work plan, design of questionnaires, conduct of interviews, analysis of the empirical study and writing the final report. In the function of an auditor, IKU could contribute its expertise on the Hungarian research landscape while the transfer of western methods of evaluation and audit were possible via ISI.

4.2.2.4   Data Collection

The most important methods of data collection were personal interviews with enterprises, research institutes and experts which were carried out from April to August 1996. Besides the direct collection of data, background information on biotechnology in the country and recent international developments were analysed as a basis for the interpretation of the results of the empirical study.

In order to create a valid data basis for analysis, considerable focus was placed on the careful preparation and conduction of the interviews. The composition of the sample, as well as of the interview team, the preparation of interview guidelines and the final documentation proved to be crucial factors. Best practice with interviews of technology oriented enterprises and research institutes in Germany and other advanced market economies had to be adapted to the Hungarian environment. The main findings and employed best practices will be described in the following.

Composition of the Interview Sample of Enterprises and Institutes

Valid and internationally comparable data are very difficult to obtain in economies in transition. Because of the restructuring of the economy and legislative changes in the field of reporting requirements, especially registers of enterprises tend to be incomplete or even outdated. In the case of biotechnology, as for other technology fields in general, the availability of complete registers is even more complicated, since the technology is applied in various industrial sectors. Therefore, in the Biotechnology Audit in Hungary, composing a sample of actors in biotechnology which was to be as complete as possible meant employing a very broad approach.

In the first approach, a list of all potential actors in biotechnology (especially enterprises) was computed using a number of registers provided by different authorities and professional organisations.[9] In addition, recent literature on biotechnology, press releases and material of conferences were analysed. Additional information emerged during first contacts and interviews with enterprises.

During first contacts many enterprises dropped out as potential actors in biotechnology, since they claimed to have moved out of the field. Especially high uncertainty about future activities in biotechnology was associated with acquisitions by foreign investors. A number of firms refused to be visited during first contacts, but could be

---

9 Registers used were: List of owners of plant varieties, R&D register at the National Technical Information Centre and Library (OMIKK), list of organisations which received grants from the Hungarian Research Fund (OTKA), list of organisations which received grants or support from the Central Technological Development Fund (KMÜFA), list of pharmaceutical firms registered by the Central Statistical Office, lists of biotechnology-related engineering and scientific professional organisations.

identified as relevant actors. It was impossible to cover the micro-sector among biotechnology firms since there is no official reporting on their activities. Given these limitations, the remaining sample could be considered as a reliable basis for drawing realistic conclusions on biotechnology from the interviews.

As was expected, the identification of research institutes and university departments was straightforward, due to officially provided data and IKU's familiarity with the Hungarian research landscape.

In order to identify the best respondent within an organisation, CEOs at firms, directors at institutes and heads of university departments were contacted. Interviews were carried out with directors of departments at institutes and universities and with top management and/or R&D responsibles in enterprises all over Hungary. Expert interviews were an important information source throughout the project. Besides full-length interviews with experts, they were extremely helpful to validate the composition of the sample and, later on, the results of the audit.

Questionnaires for Interviews

For the Biotechnology Audit in Hungary, the elaboration of a questionnaire as a guide for the interviews proved to be valuable in order to secure comparability between interviews and cover all relevant issues. To explore the interview situation close and open-ended questions were included. Quantitative data were complemented by qualitative information. The questionnaires investigated the situation and activities of the individual interviewee and asked for an assessment of other actors relevant for biotechnology in Hungary and of the overall performance in terms of technology and markets of the country. Besides more general information, indicators covering the current technological and market strengths of the interviewees and the future perspectives were employed. Special focus was on the organisation of research and of market relationships (network relations and internal management procedures) as well as on policy matters. The interviewees were invited to prepare a SWOT analysis to condense the detailed information .

The questionnaires were prepared in English and Hungarian. Often interviews took more than four hours which indicated, among others, high interest in the action on behalf of the interviewed person. Interviews were carefully documented.

Interview Team

The major part of the interviews was carried out by a mixed team in terms of nationality (German, Hungarian) and scientific background (biology, economics). This made it possible to clarify possible misunderstandings arising from language or technical problems. Only some interviews were possible only with Hungarian experts from IKU, since the interviewees refused to respond to western interviewers or could not speak other languages than Hungarian.

4.2.2.5   Data Analysis

The data analysis was carried out between August and October 1996. The analysis of the field study aimed at providing statistical, valid data which would reveal the strengths and weaknesses of biotechnology in Hungary, with special focus on the functioning of the innovation system and possible deficits.

The quantitative analysis on the micro-level was complemented by qualitative research on the history and on framework conditions of biotechnology in Hungary. This proved to be very helpful not only for the preparation of the fieldwork, but also for the interpretation of the results of the empirical study.

The assessment of the possibilities for bibliometric analysis led to the conclusion that these were not existent for biotechnology in Hungary. Biotechnology-related Hungarian journals concentrated on new scientific results rather than their application and contextual factors. Similarly, the sectoral literature survey showed a highly limited number of studies carried out.

To assess the future perspectives for biotechnology in Hungary, an intense international study covering the performance of different advanced market economies in biotechnology was carried out. The requirements for competitiveness were analysed, taking into account the specialities of the technology. This was considered as one of the key instruments to arrive at realistic results in the study. The results of the empirical study were evaluated against international developments and best practices.

4.2.2.6   Generation of Results and Recommendations

The results and recommendations of the audit have been presented to the participants of the action in a workshop held on November 21st 1996 in Budapest. The aim was to validate the results, give feedback to enterprises and institutes, bring dispersed actors together and present the policy recommendations to representatives of the Hungarian government and the technology policy administration.

Representatives from institutes, university departments, firms, the patent office, associations, ministries, the OMFB and the BMBF participated in the workshop. The workshop presented international trends in biotechnology, the Hungarian situation and pointed out bottlenecks of future development. The recommendations benefited from the international comparison of strategies pursued successfully by enterprises and institutes in biotechnology but also of alternative government support schemes aiming at technology transfer between science and industry.

The various results of the audit were supported and emphasised by the audience. Systematic deficits of the Hungarian system of innovation were discussed as well as concrete obstacles. To name only a few, the knowledge transfer from science to industry could become a crucial factor hampering the exploration of potentials in biotechnology. Enterprises and institutes expressed their need of improved information on financial support on behalf of the technology policy administration and of a more effective enforcement of the patent legislation. Many of the participants expressed to have learnt new information which would be valuable for their future strategies.

The audit identified as a priority area for biotechnology the agro-food sector in which Hungary possesses technological strengths and has chances to explore its existing potentials. This priority was discussed and validated during the workshop.

4.2.2.7   Dissemination and Implementation

After the Biotechnology Audit, the implementing of policies depend on the Hungarian government. The audit has identified areas calling for action, in which policy measures could be elaborated together with the OMFB. Biotechnology is an important issue in Hungarian technology and innovation policy. The information and priorities provided by the audit will serve as input for political decisions of the OMFB and the government. At present the implementation of concrete policy measures is not clear.

One of the possible follow-up activities recommended was an international workshop to further disseminate the audit results in biotechnology. This could serve to convey findings about the innovation system in Hungary to a broader public within the country as well as in other CEECs. Furthermore, international scientific cooperation could focus on and support financially the identified technological strengths of Hungarian biotechnology.

5.   Guideline for Technology Audit in CEECs

5.1   Introduction

The preceding chapters have explored various sources of alternative methodologies for technology audit: a broad literature survey and the evaluation of two pilot technology audits in Hungary have been undertaken. Chapter 5 aims at the synthesis of the research and presents Guidelines for Technology Audit in CEECs as international best practice. A methodology is selected which meets the necessary requirements given the constraints in transition economies: availability of data, validity of data and of results and their contribution to the identification of technological strengths and weaknesses and the formulation of a strategy for technology policy.

The guidelines offer different modules for alternative design of a concrete audit project and its needs. The audit process itself and the inventory of possible methods and indicators may differ with different overall audit objectives. Examples are given below to illustrate this.

5.2   Audit Objectives

The overall rationale of technology audit has been discussed in detail in Chapter 3. For a concrete audit plan, there are different possibilities to design the audit, according to the objectives and needs of the country in which the audit is to be carried out. Responsibility for the action has to be assumed by the national government or by public authorities for the following reasons: the first reason is political in nature, since the audit may interfere with internal interests of the country. The other reasons are practical in nature, because the success of the audit depends on official support during the fieldwork and the implementation of recommendations. Therefore, it is the task of the government (or their representatives) to define the objectives of the audit, i.e. the target groups for the results of the audit and the subjects of investigation. The main issues of the audit can be defined through background research or a mere political vote considering specific interests or a combination of both. The phase of initiating a technology audit is illustrated below (figure 5.2-1):

Figure 5.2-1:   Defining the Objectives of Technology Audit in a Country

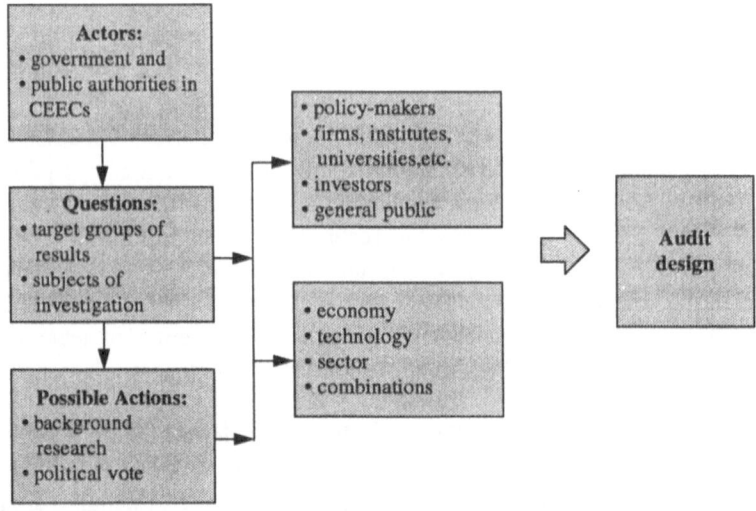

Target Group of Audit Results

The target group of the audit is to some extent interdependent with the selection of subjects for investigation and their aggregation level. The following target groups may be addressed by the result of the audit with different priority: policy-makers, actors involved such as enterprises, institutes, universities, intermediate institutions, investors and the general public. Consequently, the results which are expected of the audit differ. For example, if government is the main target group, the main emphasis will be on recommendations for government action such as policy measures. Where policy recommendations are requested, it has to be clarified which level of concreteness is necessary and which is the scope of the policy fields involved. If firms are the main target group, consulting and enabling activities may be in the foreground.

Subjects of Investigation and Level of Aggregation

To map the whole range, alternative subjects of investigation can be the whole economy, specific technologies, selected industrial sectors, a region or a combination of these. These approaches are of varying usefulness. The evaluation of technological strengths and weaknesses of the **whole economy** seems impossible, and therefore can be excluded from the further discussion. As the first pilot audit in Hungary has shown, the focus on selected **sectors** of a national economy can generate valuable insights. Nevertheless, for the evaluation of future technological perspectives, a **technology**-oriented audit seems more promising. As new technologies can be cost-saving devices in industry and can offer new production possibilities, the technological capabilities of a country shape the competitiveness of various sectors of the national economy. Technological capabilities consist in scientific research, technical skills to further develop and apply new techniques and marketing know-how, as well as the infrastructure and environment conditions. There are alternative criteria to depict the technology field: either the audit focuses on the technological and industrial strengths of the country, or selected areas which are associated with high potential but also high risk. The evaluation of technological strengths of a region has to take into account both the sectors in the region and the technologies employed. The analysis could pursue a combined approach, identifying main specialisation patterns and assessing them in a national and international context.

Besides the broad definition of the subjects of investigation, as suggested above, further specification is needed. For example, a technology audit can be used as an ex-ante research tool for designing technology policy programmes, restructuring public research or the internal organisation of research institutes, or can point out technological areas of interest for potential investors. In these cases, the subjects of investigation and their aggregation level will differ. The research questions of the audit can be derived from the underlying logic of the examined sub-system through background research. This especially is true for the specialities of a technology and modern industrial organisation.

5.3   Selection of Auditors

> **Auditors:**
> - independent
> - national and foreign
> - technical and economic expertise
> - language skills
>
> *optional:* involvement of government representatives

The selection of the auditors is a crucial issue for the whole audit. Summarising the experiences made during the pilot actions, they determine, among others, standards of organisation of the audit, validity of information gathered during fieldwork, points of reference for interpretation of audit findings and side effects such as interactive learning among auditors. Especially auditors' independence is necessary for the confidence of interviewees. The teamwork of national and foreign auditors as well as of experts with different scientific backgrounds (i.e. technical and social/political/economic scientists) contributes to the careful interpretation of empirical information but also allows the transfer of international best practice to the CEECs. Concerning the appointment of international auditors, specialisation and international leadership in the respective field and, at best, previous experiences in CEECs are of extremely high value.

It seems to be appropriate to examine the specific attitude towards foreigners in the country, especially since the role of westerners may be difficult. Alternative funding possibilities of the audit may cause a pre-selection of e.g. nationality of auditors, if international donors are involved.

Arguments exist in favour of, but also against the active involvement of the government of the respective country. If the government is not actively involved in the audit process, this will underline the independence of the auditors not only from business interests but also from political interests, which could be an advantage in interviews. But at the same time, an active involvement of public authorities is absolutely necessary at the stage of defining the main issues of the audit and consequently during the generation of viable and politically supported recommendations.

5.4   Data Collection

5.4.1   Interdependence between Selection of Indicators, Data Sources and Methods of Analysis

There is a high interdependence between the selection of indicators, available data sources and resulting methods of analysis. This will be illustrated by the following examples. The indicator selected is "R&D strategy". Consequently, the only possible data collection method are inquiries at the micro-level of enterprises or research institutes: while postal questionnaires do not necessarily generate valid answers, interviews proved to be an appropriate tool. The analysis of the indicator is qualitative in nature.

The indicators chosen are "publications, citations and patents" (cf. figure 5.4-1). In advanced market economies, patent data banks and international citation indices provide the information required for quantitative bibliometrical and technometrical analysis to identify the technological specialisation of a country. On the contrary, in the case of CEECs, the non-availability of these data sources prescribes different methods of data collection and analysis. Information on patents, publications and citations may be collected on the micro-level and indicate, in rare cases, the scientific and technological performance of an industry or the whole country, but can be analysed qualitatively in terms of enterprises' and institutes' competitive position and strategy.

Figure 5.4-1:   Interdependence between Indicators and Methods of Analysis

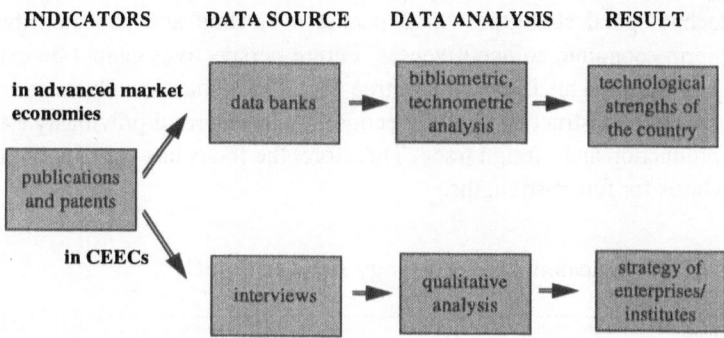

Keeping in mind the strong interdependence between availability of data sources and selection of indicators which again determine the possibilities of analysis, the following sections comment the issues separately.

5.4.2   Selection of Indicators

Categories of Indicators

The inventory of indicators which can be employed in technology audits can be broadly categorised according to their level of aggregation into micro and macro indicators, but also into quantitative and qualitative indicators.

Ideally, indicators referring to the macro-level of analysis support the interpretation of findings on the micro-level. Furthermore, quantitative and qualitative indicators should be used complementarily to provide a more complete and rich picture of the situation. Especially, many properties which constitute innovation potentials are not tangible and not subject to statistical reporting activities. However, these rationales on how to use different types of indicators become more complicated in CEECs. To illustrate this, quantitative macro indicators are difficult to obtain: either statistics do not exist, definitions often are not compatible, or if the data were available in principle it cannot be retrieved at the aggregation level needed, e.g. the respective technology field or sector. As a result, empirical research on the micro-level has to compensate for the lack of appropriate data on the macro-level.

How to Use the Inventory of Indicators

The figure below presents four types of indicators: quantitative/macro, quantitative/micro, qualitative/macro and qualitative/micro. There is a further distinction made between information relating to the overall characterisation of the situation, to existing technological assets and to strategic assets. These categories comprise both static and dynamic indicators. The key question of technology audit is the identification of technological strengths which persist over time and can contribute to found long-term economic competitiveness. Future perspectives cannot be extrapolated from past trends. This is even more true for CEECs than for advanced market economies, since the restructuring of the economy has distorted previously existing patterns in production and foreign trade. Therefore, the focus has to be on indicators which are a basis for future strengths.

Figure 5.4-2:   Indicators for Technology Audit in CEECs

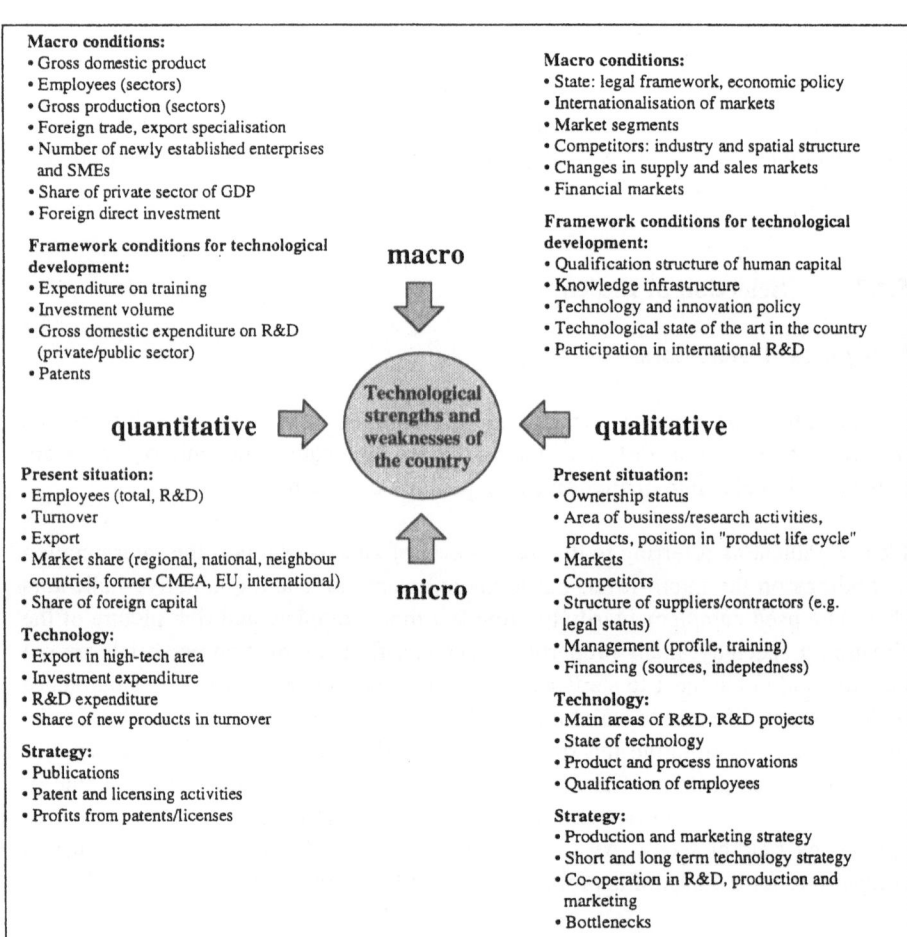

Each of these categories should be employed in the technology audit. The intensity depends on the objectives of the audit project: on the target group, the subjects of investigation and country-specific factors. Examples will be given in the following sections. On the one side this may lead to the extension of the range of indicators relating to a specific research question. On the other side, from the indicators presented in the following, a certain number should be selected. In many cases it will not be feasible to obtain information on all subjects and might be even contraproductive. Then the focus on the core questions of the audit seems necessary to obtain data which is valid and can be interpreted. A good tool to assist in the selection of indicators is a pretest with a small sample of respondents and experts.

Quantitative Indicators on the Macro Level

Quantitative data referring to the overall performance and structure of the economy are: GDP, production output (sectoral), employees (sectoral), foreign trade balance. The dynamics of the economy can be measured by the number of newly established businesses, SMEs and the share of the private sector of GDP. Expenditure on training, investment volume, Gross Domestic Expenditure on R&D (private and public sector), domestic and patents granted abroad convey a first impression of the framework conditions for technology development. Volume of FDI can be used as a measure of the ability of specific sectors to attract external capital.

Qualitative Indicators on the Macro-Level

Qualitative information complement the quantitative data: legal framework (property rights, tax and foreign trade regulations, company law), administrative efficiency, underlying institutional changes. Quantitative data should only be interpreted taking into account the underlying institutional changes which have had an impact in the recent past or will affect quantitative indicators in the future. Framework conditions relating to technology are: qualification structure of human capital, the knowledge infrastructure including universities, research institutes, technology transfer institutions, business associations, chambers etc. and technology and innovation policy, technological standards and integration in the international scientific community. Technology policy includes institutional funding of the research infrastructure, government support schemes, funds and associated budgets for science and technology in the public sector and in industry.

Possible starting points for **further investigation** could be: the political environment, the functioning of the knowledge infrastructure, the structure, dynamics and needs of the business sector, financial markets and harmonisation requirements in the context of trade or association agreements with the EU. Emphasis will arise from the objectives of the audit. Depending on the subjects of investigation of the audit, which for example can be a key technology or an industrial sector employing lower technologies, the focus of analysis will shift between more technology-related and more business-environment-related indicators.

Quantitative Indicators on the Micro-Level

General characteristics of enterprises or research institutes, universities or other relevant actors are number of employees, structure of qualification, turnover, surplus, rentability, exports, market share (regional, national, neighbour countries, former CMEA, EU and international), capital structure, especially foreign capital. Quantitative input indicators relating to the state of the technology are: expenditure on investment and R&D expenditure. They are preconditions for future technological development. The export in high tech areas and the share of new products indicate past achievements and the ability to transform technological strengths into market success. Publications refer to scientific results and can measure the performance of science in the pre-competitive or basic research. Patents and licenses reveal the possible market income which is associated with an innovation.

Qualitative Indicators on the Micro-Level

Given a fast changing environment, qualitative indicators relating to the process of production and innovation will be one major source for the information necessary to evaluate technological strengths and future potentials of industries. Indicators on the overall situation are: age of the organisation, owner status, qualification and fluctuation of employees, area of business or research activities (for research institutes), product/research portfolio, degree of diversification, position of products in the "product life cycle", organisation including hierarchies, markets, competitors, structure of suppliers and customers, management (e.g. profile of entrepreneur, training of employees) and financing. Indicators relating to markets are relevant both for enterprises and research institutes: answers reveal possibilities and potential barriers of the transfer of research results into industrial application.

The indicators should take into account the specialities of the business sector in CEECs: for example, not only the structure of suppliers and customers is relevant but also their legal status as private or public firms. This determines the market conditions of individual enterprises. Also indebtedness is a problem specific to restructured enterprises in CEECs. Additional indicators can be derived through background research on the specific situation of the respective country.

Indicators referring to technological assets are: main areas of R&D, R&D projects and their maturity, state of technology, product and process innovations and qualification of employees. These indicators characterise the starting point for technological development. The business and technology strategy indicates the actor's ability to adapt to the requirements of a market economy. Alternative basic technology strategies are: technology lead, niche strategy, "follower" strategy. To achieve these aims, different R&D strategies can be pursued: internal development, external acquisition of new technologies, patenting or secrecy of results. Besides these strategies, the processes within the organisation are important: how are decisions made and implemented in the organisation. The ability to compute and make use of various internal and external flows of information is a necessary input for flexible reaction to new technological and market opportunities. This is associated with different internal organisation structures but also various forms of co-operation with research or market partners. Indicators referring to quality are internal controlling systems, rate of rejections in production and compliance with contracted dates.

If the priority target group was individual actors such as firms and research institutes, emphasis would be on different management tools and organisation structure within and outside the borders of the entity.

The selection of qualitative indicators on the micro-level has to be aware of the constraints associated with the stage of development and formalisation of enterprises and institutes. A warning relates especially to indicators such as the internal controlling system or management tools employed. They can only be employed after assessing the overall situation and development stage of respondents - in many CEECs this will not be possible.

5.4.3   Data Sources

Empirical research should be carried out at two phases during the audit, first as background research and secondly, during the core of the audit, which is the fieldwork. The table below presents a suggestion for different data collection methods which are believed to suit the research questions identified for each step best. They will be discussed in the following sections.

With reference to the organisation of the audit process, background research has two functions. On the one side, it provides input for the definition of the main issues of the audit through a rough analysis of possible technological strengths of the country. On the other side, background research is a necessary preparation for the fieldwork to assess data availability, clarify the specialities of the technology or the sector in the country and the relevant context factors.

Table 5.4-1:   Stages of Empirical Research

Step 1: Background Research	
(1) research questions and objectives	• selection of subjects of investigation • availability of data • specialities of sector/technology • context factors
(2) data collection methods and data sources	• survey of literature (statistics, journals, studies) • indirect data collection through expert interviews • expertise/experience of auditors
Step 2: Fieldwork	
(1) research questions and objectives	• representativeness of sample • validity of data • dynamic analysis (future perspectives)
(2) data collection methods and data sources	• interviews • postal questionnaires • expert workshops • case studies • survey of literature (statistics, journals, studies)

In principle, the data collection methods for technology audit in CEECs are to some extent similar to methods employed in advanced market economies. This has been indicated by the literature survey (cf. table 3.2-1 and table 3.3-1) and was a result of the pilot actions. Nevertheless, in the case of CEECs, there are some methods which can only be fruitfully used when overcoming associated constraints and there is even a group of methods whose use is highly limited or even unrealistic:

Transferability of Data Collection Methods to CEECs :

(1) **high:**
  ⇨ direct and indirect data collection in general
  ⇨ expert interviews
  ⇨ expert workshops
  ⇨ visit to fairs
  ⇨ case studies

(2) **with constraints:**
  ⇨ exploitation of existing data (statistics, studies, journals)
  ⇨ postal questionnaires
  ⇨ interviews
  ⇨ representative samples and control group analysis

(3) **with severe constraints:**
  ⇨ panels
  ⇨ data banks

In general, direct and indirect data collection methods are fruitful for technology audit in CEECs and it is recommended to use them complementarily.

Indirect data collection through **interviews and workshops with experts** signifies a first approach towards a comprehensive and consistent representation of the audit subject during the background research. They may indicate critical or complementary views during the fieldwork stage of the audit. Possible topics for expert consultation are: exploitable data sources, framework conditions, recent political changes, mapping of main players, identification of main problem areas, design of questionnaires, validation of audit results, and generation of policy recommendations. As the pilot audits in Hungary have shown, it is crucial to find the appropriate person who is able, willing and as objective as possible to evaluate the situation.

The **visit to fairs** represents a more random approach to map the actors' landscape in the country, though they are an opportunity to conduct open and direct discussions and interviews or to establish first contacts.

Case studies provide detailed insights into e.g. the organisational structure of an institution or the logic of decision-making processes. They can be employed to highlight main characteristics which are found to be general properties.

While the identification of the technological position of a country can rely on **panel data and data banks** in advanced market economies this is not the case in CEECs. Panel data and data banks are rarely available and in most cases, not tailored to the needs of technology audit. Therefore, it does not seem possible to found main analysis on these sources. Nevertheless, wherever data banks exist, they should be explored as much as possible.

The **exploration of existing data** is recommended as highly valuable. Still, official statistics can only be used to very limited extent, though they should be explored especially on the micro-level as an input for the empirical study. Statistics are only starting to be harmonised with international reporting procedures and mostly are only available with some delay. In some CEECs, the number of studies being carried out recently is very promising. Since the beginning of the transition period, research including empirical studies on issues relevant for technology audit has increased considerably (cf. survey of literature). It is necessary to assess the country-specific situation to identify all possible sources for the audit.

Detailed Description: Interviews and Postal Questionnaires

The methods which should be utilised in the empirical investigations during the technology audit are interviews and postal questionnaires. Although they have been proved capable of generating fairly representative and valid data, they are associated with certain constraints which have to be overcome. In view of the high relevance of these instruments, details are given in the following.

Composition of the Sample:

(1) **Assessment of data sources**: possible sources differ greatly according to countries: registers of public authorities such as ministries, funds, clearing houses and of private organisations, data banks, journals, daily press, experts, field study.

(2) **Scope of field study:** depending on the size of the country and the focus of the audit, the fieldwork can attempt either to cover all relevant actors or select a representative sample. In the first case, after the computation of a preliminary list of actors which is as exhaustive as possible the elimination of non-relevant elements takes place during the audit itself. But also the selection of a representative sample requires knowledge about the composition of the total population.

(3) **Selecting a representative sample:** the sample should have the same structure as the total population in terms of key characteristics (e.g. size in turnover and employees, age, activity, R&D intensity). During the empirical investigations, actors will entirely drop out of the sample or show different characteristics than previously assumed. In these cases, a comparable substitute should fill the gap and readjustments have to be made. The computation of a representative sample depends greatly on the knowledge of the basic population. For some CEECs, the possibility of conducting a representative selection seems highly *unrealistic*.

In some countries, it seems to be realistic neither to conduct an exhaustive nor a representative field study. The selection of actors for empirical analysis then has to rely on more vague assumptions about the overall population. Technology audit is still possible, but the results generated have to be interpreted very carefully.

Interviews:

(1) **Scope of interviews:** interviews have proved to be a necessary complement in transition economies and may even substitute postal inquiries. Costs associated with interviews are considerably high.

(2) **Interview team:** the interview team should consist of mixed expertise (technical and economic/social/political scientists) and involve members of the respective CEECs who speak the language(s) of the country but are also fluent in the communication language (e.g. English).

(3) **Preparation of interviews and documentation:** to assure comprehensive and comparable data collection, the preparation of an interview guide is highly recommended. This should be available in the language(s) of the country. In addition, interviews should be documented carefully.

(4) **Establishing the contact:** to target the appropriate person within the organisation is of importance because of responsibility and information issues.

(5) **Validity of answers:** interviews allow for the clarification of possible misunderstandings. At the same time, the auditor may ask additional questions exploring the direct contact and is able to gather further information through visiting the site or receiving information material. This allows for a more realistic estimation of the actors' situation.

(6) **Type of indicators:** benefiting from the interactive situation during the interview, the auditors may use more explorative types of indicators which have to be explained to the interviewee.

Postal questionnaires represent an additional instrument for empirical investigation.

Optional: Postal Questionnaires:

(1) **Cost advantage:** postal questionnaires represent the opportunity to cover a larger number of respondents with relatively low costs by comparison to on-the-spot interviews.

(2) **Establishing the contact:** finding the appropriate person in terms of responsibility, access to information and skills to answer the questionnaire is equally important to convincing them to participate in the survey. A recommendation on behalf of the government can be helpful.

(3) **Clarity of questions:** clearly stated questions are a prerequisite for valid answers. In advance, auditors have to avoid possible misunderstandings. In many cases, approaches may still be dominated by the socialist legacy.

(4) **Language:** questionnaires have to be in the national language(s).

(5) **Hotline:** a lot of emerging questions will be solved, if a hotline is provided on behalf of the auditors.

(6) **Validity of answers:** as the pilot actions have shown, self-evaluations may be estimated over-positively. Therefore, in any case, postal questionnaires should be complemented by a whole interview round or at minimum by random on-the-spot visits.

(7) **Representativeness of answering actors:** in most cases, answering the questionnaires is not compulsory, therefore reminding action through sending new questionnaire, telephoning, faxing or even telephone interviews can be employed.

(8) **Type of indicators:** in general, open and closed questions are feasible, though closed questions with standardised answers are highly recommended as they significantly facilitate the analysis later on.

5.5   Data Analysis

A wide range of possible methods for data analysis exists, which represent contemporary research on technological potentials or have been employed during the pilot actions in Hungary. Their applicability for technology audit in CEECs is illustrated by the rough estimation below:

Transferability of Data Analysis Methods to CEECs :

(1) **high:**
- ⇨ qualitative
- ⇨ qualitative: e.g. SWOT analysis
- ⇨ quantitative: statistical

optional:

(2) **with constraints:**
- ⇨ bibliometric analysis
- ⇨ patent analysis
- ⇨ Delphi method

(3) **with severe constraints:**
- ⇨ technometric analysis
- ⇨ econometric models
- ⇨ simulation analysis

Technology audit primarily relies on more common quantitative and qualitative methods of data analysis, i.e. the identification of statistically significant patterns and correlation between key characteristics. The recommendation is to complement quantitative with qualitative data analysis.

A good tool to conceptualise the findings of the audit is the **SWOT analysis** which can be compiled by the individual actors as well as on aggregate level. Abstracting from detailed and concrete questions, it identifies main internal and external factors which are crucial for future success. Besides the balance between internal strengths and weaknesses, the confrontation of internal and external determinants reveals the fit or need for adaptation. These qualitative issues can be the starting point for recommendations.

Figure 5.5-1:   SWOT Analysis

INTERNAL ANALYSIS	**STRENGTHS** keep, reinforce	**WEAKNESSES** improve
EXTERNAL ANALYSIS	**OPPORTUNITIES** exploit	**THREATS** eliminate, adapt

(Based on European Commission 1995, p. 261)

A crucial issue is to derive a representative picture of the audit subjects from the empirical analysis. There are two problems: on the one side, individual answers still may not be valid in a very strict sense, e.g. having a systematic bias because of the respondents' limited perception. On the other side, the aggregation of the results will be difficult if the statistical representativeness of the sample cannot be guaranteed - which might even apply to the majority of studies in CEECs.

Therefore, interviews or postal questionnaires should have internal cross-checks which allow for the assessment of the robustness of subjective statements. Still, the adjustments of respondents' answers by auditors is very delicate and should be avoided if possible. A more feasible approach is the confrontation of the results of the audit with quantitative or qualitative macro-data such as official statistics or experts' evaluations wherever available and appropriate.

Possible **points of reference** for evaluation of the observed technological strengths and weaknesses can be: international developments and best practice, performance of other CEECs with comparable position. This has to be judged according to the technological and market characteristics of the respective technology or sector, e.g. competition may be global or regional. If a key technology is investigated the points of reference will be international leading developments. On the contrary, methods of technology foresight and Delphi methodology seem only be applicable to a very limited extent as they represent a very sophisticated methodology which might not help to solve the main issues of the analysis. During the pilot actions, Delphi methodology has been employed to generate a consistent evaluation of the present situation in CEECs through expert statements. While it might be an option to generate an aggregate picture, this indirect method of data collection cannot compensate for empirical analysis.

The following methods can provide results on an aggregate level but are associated with difficulties in CEECs: bibliometric analysis, patent analysis and technometric analysis. Bibliometric analysis depends largely on the availability of national or international journals in which research results are published and whether the classifications available are compatible with the specific questions of the audit. They appear to be a useful tool to approximate the main areas of research of CEECs, and their strengths and weaknesses during the stage of background research. However, this picture will not provide any indication of actual transfer from science into industrial application.

Patent analysis indicates the performance of applied research in science and industry. Systematic patent research as carried out in advanced market economies is not possible. The applicability of patent analysis is similar to bibliometric methods: harmonisation of patent classification with international standards is only in its infancy and past national patents are only of little comparability. Patents granted by the European patent office or in the US may be a help, but this is an indicator which applies only to a part of CEECs. Furthermore, classifications may not be tailored to the needs of the audit. Patent data may be used as a very broad approach.

On the contrary, econometric models and simulation analysis do not seem to be fruitful methods for technology audit. They signify an over-simplification of the complex situation.

5.6 Generation of Results and Recommendations

In assessing the technological strengths and weaknesses of a country the aim of technology audit is to identify competitive areas with potential for future development. The evaluation of recent performance in the light of future challenges leads to recognising, stimulating but also restraining factors. In the "Common Methodological Principles" for the sectoral audits in Hungary (cf. Annex I) the OECD named some general and sector-specific restraining and stimulating factors such as lack of qualified suppliers, underemployment, standards, business environment and management issues. Restraining and stimulating factors can constitute the starting point for the design of recommendations. Stimulating factors need further developing while restraining factors have to be eliminated. Recommendations may be related also to long-term goals, but the primary aim should be to provide concrete measures.

After the identification of existing deficits, possible actions as well as responsible actors should be named. In many cases, transparency of all parties which are involved or affected by policy measures does not exist and should be provided by the audit. Such an overview would facilitate the necessary co-ordination between the actors. As the audit provides a unique collection of most recent and reliable data, the action should contribute to the formulation of an overall technology policy strategy pointing out areas with particular technological strengths which should be further developed. Therefore the audit plays an important role in the setting of national priorities for the allocation of scarce resources in science and technology. The audit results may touch upon critical and delicate issues of national interest and even contradict former perceptions of national competitiveness.

Recommendations as a result of the technology audit depend strongly on the objectives and target groups of the audit defined in advance.

Final results and recommendations rely on the quantitative as well as qualitative analysis of the background research and the empirical analysis. Technological strengths can be identified, benefiting mostly from the expertise of the auditors and from international comparisons. Alternative scenarios can be helpful, since predictions and precise formulation of future perspectives seem highly difficult, especially in the transformation context. To generate recommendations the auditors should involve outside experts and policy-makers in order to concretise and validate their advice. Moreover, the delicacy of politically targeted technology or sector policies requires the responsibility of political actors and cannot be assumed by the auditors.

Table 5.6-1:   Target Groups and Recommendations of the Audit

Priority target group	Areas of possible recommendations
Policy-makers	• macroeconomics (e.g. monetary and foreign exchange policy) • institutional aspects (e.g. property rights, tax legislation, company law, banking regulations) • reorganisation of research institutes • foundation of institutions of the knowledge infrastructure for technology transfer • implementation or reorganisation of technology programmes: focus and components • overall perspectives for state enterprises
Individual actors: enterprises, research institutes, intermediary institutions	• action plans for strategic planning, organisation, financing, marketing • formulation of technology strategy • dissemination of international best practice

5.7   Dissemination and Implementation of Audit Results

Recalling one of the main aims of technology audit which is to provide a sound basis for political strategy formulation in innovation policy, audit results are addressed to governmental officials and public bodies. In addition, the audit represents an important opportunity for CEEC policy-makers, experts, institutes and enterprises to participate in a learning process.[10] Therefore, dissemination strategy has to target these groups.

Follow-up activities to disseminate the results of the audit may be workshops, conferences, publications and consulting activities to policy-makers and enterprises. The implementation of the results can be through action plans and pilot projects. For instance, possible measures can be training activities and the implementation of new institutions or the establishment of a network between already existing entities. The identified areas with technological strengths can be selected for bilateral or multilateral co-operation.

The implementation depends largely on the interest shown by these groups and the financial resources available. Therefore, it is recommended to allocate a separate budget for follow-up activities right at the beginning of the audit.

5.8   Concluding Remarks

The report presents Guidelines for Technology Audit in CEECs to national governments who are interested in such an action and to researchers who will carry out the analysis of technological strengths and weaknesses. The researcher will find at his/her disposal a guideline on how to conduct the audit process, which instruments can be used and which are the criteria and actions to select the audit design. The methodology of technology audit has been formulated on the basis of theoretical analysis and empirical work. The experiences of the Hungarian Pilot actions have validated the applicability to CEECs.

The guidelines of technology audit describe in detail the elements and alternative research methods which can be employed, but emphasise also the organisational requirements of the audit process. The suggested framework does not represent a fixed prescription. Instead, basic elements and options for further investigations are given. Furthermore, alternative foci which arise according to shifts in the objectives

---

[10] The OECD Methodology of Audit (1995a) asks to involve appropriate government institutions, firms, institutes and related institutions, national experts, technical and management expertise, although, it is not entirely clear whether these actors should be subject to audit or should be actively involved in carrying out the audit. For further details, see Holland (1995).

of the audit are described. The design of a specific audit project in a CEEC will depend on the objectives formulated by its government and the specific socio-economic and technical conditions.

Technology audit is an instrument which provides governments in CEECs with reliable information as a basis for policy formulation and assists in the economic and technological reorientation of transition economies. Auditors can be regarded as external catalysts to the development process. In order to cope with the structural discontinuities, the methodology is designed not only to assess recent performance but also to detect technological potentials. This objective causes a considerable challenge to the researcher being stressed by the existing difference in socio-economic background and perception in CEECs which at any stage of the audit has to be considered.

II.   Biotechnology Audit of Hungary

Analysis of Strengths and Weaknesses of Biotechnology
in Selected Sections

1.   Introduction

In its review "Science, Technology and Innovation Policy Hungary" (1993) the OECD recommended a technology audit for the country. As the Hungarian authorities were highly interested in the project, the OECD carried out a pilot technology audit in co-operation with research institutes from Germany, Finland, France and Austria from April until December 1994. Its aim was to analyse industrial and technological strengths and weaknesses in order to provide a sound basis for technology policy. The National Committee for Technological Development (OMFB) in Hungary selected four sectors which were considered to have potentially competitive strengths: medical equipment production, packaging industry, agricultural machine production, and plastic processing industry.

The Biotechnology Audit for Hungary is based on the results and experiences of this preceding audit. It was carried out within the framework for Technical Co-operation of the German Federal Ministry of Education, Science, Research and Technology (BMBF) for Hungary. It is a close co-operation between the German Fraunhofer Institute for Systems and Innovation Research (FhG ISI) and the Hungarian Institute for Innovation Research (IKU). The aim is to assess a technology field instead of a specific sector, since modern technologies tend to alter the borderlines and links between traditional industrial sectors. Besides its generally perceived high potential as one of the critical technologies for the 21st century, biotechnology was chosen on behalf of the Hungarian authorities as Hungary is assumed to have specific strengths in this area.

Besides the potential of biotechnology to provide new products, processes and services, two additional characteristics of biotechnology are mainly crucial for this assessment: firstly, biotechnology is a universal enabling technology which can be used within different industrial sectors at different innovation stages. At present this may already be seen within the pharmaceutical industry, where the development of new drugs is almost impossible without using biotechnological methods. Secondly, modern biotechnology is considered as a basic technological precondition for the further interdisciplinary development of other critical technologies. This means that numerous important areas of technologies will be influenced by biotechnological approaches. An example of this feature is the relation between biotechnology and information technology, where a new interface between these different technologies has been developing recently: the field of bioinformatics. In general this increasing intertwining of different areas of technology will become an essential feature of the technological development at the beginning of the 21st century (Grupp 1993).

For this project a rather broad understanding of biotechnology is used defining biotechnology as any technique that uses living organisms or parts thereof to make or modify products, to degrade substances, to modify living organisms (plants, animals, microorganisms) for specific uses, or for services (e.g. in analytical laboratories). Following this definition genetic engineering is not synonymous with biotechnology but rather one of several methods which are used in biotechnology.

Modern biotechnology emerged in the early 1970s and was driven mainly through research and development in the United States. At the same time the first Hungarian research centre devoted to biotechnology was established. Its foundation accidentally coincided with the scientific revolution that resulted in the birth and worldwide growth of modern biotechnology[11].

It is supposed that Hungary has considerable academic strength in the life sciences and the question may be posed, whether and to which extent this strength has begun to be exploited for the commercial development of biotechnology.

The evolution of biotechnology in Hungary is linked with particular problems whose impacts need to be taken into consideration during this analysis. Among others, the transformation of the whole economy, radical changes in the environment and markets, the collapse of the CMEA market must be mentioned. A general limitation in financial resources also creates obstacles, not only for the effective promotion of biotechnology. Furthermore, considerable changes have taken place towards technology and innovation policy during the transition period.

Against this background the Biotechnology Audit aims to produce the following:

- an analysis of industrial and technological strengths and weaknesses of biotechnology in Hungary in an international comparison;

- an analysis of framework conditions for biotechnology in Hungary;

- deductions for technology policy options and recommendations for the OMFB.

A mix of different methods has been applied in carrying out the Biotechnology Audit:

- evaluation of scientific literature and research studies;

- utilisation of scientific, technical and economic indicators;

- interviews with experts from research institutions, firms, governmental bodies and other institutions;

- discussion of the main findings of the audit with participating institutes, firms, public bodies and policy-makers during a workshop in Budapest.

Details of these methods are described in the respective chapters.

---

[11] It is worth mentioning that the first person to use the term of biotechnology was Hungarian-born Károly Ereky who introduced this expression in his book *"Biotechnology of meat, fat and milk production in agricultural farming on a large scale"* published in German, in 1919.

In chapter 2 a brief overview of the state of biotechnology in selected countries is given. This analysis of the international situation in biotechnology will be used as a reference for the analysis of the Hungarian situation. Chapter 3, which is the core part of the study, describes the empirical analysis of biotechnology in Hungary. In chapter 4 the Hungarian situation in biotechnology will be compared with the international situation. Chapter 5 contains policy recommendations and summarises the conclusions.

2.   International Development of Biotechnology

In this chapter a brief review of the state of biotechnology in selected countries will be given. The analysis will concentrate on the United States, Japan, Germany, the United Kingdom and Israel. These countries have been chosen mainly for the following reasons: in order to derive key factors which contribute to the successful development of commercial biotechnology, it is decisive to take into account the international experience of countries which are considered to belong to the forerunners. In other words: biotechnology is a global "business" where international trends are important. Israel has been included because it is a small country which is supposed to face some problems which are comparable to the Hungarian situation.

In order to describe the situation of biotechnology in these countries the following issues will be considered:

- introduction, pointing to the significance of biotechnology within the respective country,
- research institutions in biotechnology,
- commercial activities,
- framework conditions like policy measures, legal framework, financing conditions etc.,
- conclusions.

A systematic in-depth analysis of all these issues in each country would reach far beyond the scope of this study. Therefore, the individual country analyses will concentrate on selected aspects.

2.1   USA

2.1.1   Introduction

The development of biotechnology during the past 20 years has been driven mainly by research and industrial activities in the United States. This is very obvious if, for example, patent applications in biotechnology are analysed (Schmoch et al. 1992). At the beginning of the 1980s about one third of all biotechnology patent applications at the European Patent Office originated from the United States. At the beginning of the 1990s this share increased to about 50 %.

This leading role of the United States is even more pronounced in important subareas of biotechnology like genetic engineering methods or biopharmaceuticals.

Certain factors contributed to the successful development of biotechnology in the USA. First of all the broad and highly qualified research base, especially in bio-medical sciences, provided the knowledge base for commercial exploitations. Among others the National Institutes of Health (NIH) became one of the driving forces of biotechnological research. In addition, there are a huge number of private and public universities engaged in research with high relevance for biotechnology. These research activities have been funded properly mainly through public sources. A second important factor was that the commercial potential of biotechnology was recognised very early. First cloning of a gene has been reported in 1973 (Cohen et al. 1973), only three years later, in 1976, the first biotech company, Genentech, was founded by the scientist Herbert Boyer and the venture capitalist Robert Swanson. This liaison between entrepreneurship and high quality research led to the foundation of many start-up ventures dedicated to biotechnology. This means that many scientists became entrepreneurs and were willing to run the risk of starting their own business. Only few of these later became successful, indeed many failed, but, and this is an important difference between the early days of biotechnology in the United States and certain European countries, as for example Germany, they tried to start a business again taking advantage of a climate which supports risky commercial activities. Especially the financing environment was in favour of such entrepreneurship. There was enough private capital available support to start-up companies.

2.1.2   Commercial Activities

According to different estimates, there are presently between 1,050 (Bullock and Dibner 1995) and 1,310 (BIO 1996) biotechnology firms in the United States which employ about 100,000 people (BIO 1996). "Biotechnology firm" means that the main business of these companies is biotechnology. Sometimes these companies are also called "dedicated biotech firms (DBF)". The total sales of these companies has been estimated at about 8 billion US$ in 1994 (Ernst & Young 1994a). R&D expenditure for the entire biotechnology industry in 1994 reached about 7 billion US$ which corresponds to an average R&D intensity of about 90 % (R&D expenditure as a percentage of sales). The pharmaceutical industry by comparison has R&D intensities of about 16 %. R&D expenditure per employee in the biotechnology industry are also very high: 68.000 US$ compared to 39.000 US$ for the pharmaceutical industry. All in all the biotechnology industry is one of the most R&D-intensive industries in the United States. The top 7 US companies in R&D spending per employee are biotechnology companies (BIO 1996).

The American biotechnology industry is young. Most of the companies have been founded during the second half of the 1980s, with a peak in 1987, where 121 new companies were founded (BIO 1996). Since the beginning of the 1990s the number of start-up formations has been decreasing. During 1994 the industry grew by 39 companies.

Figure 2.1-1:   Market Segments of US Biotech Firms

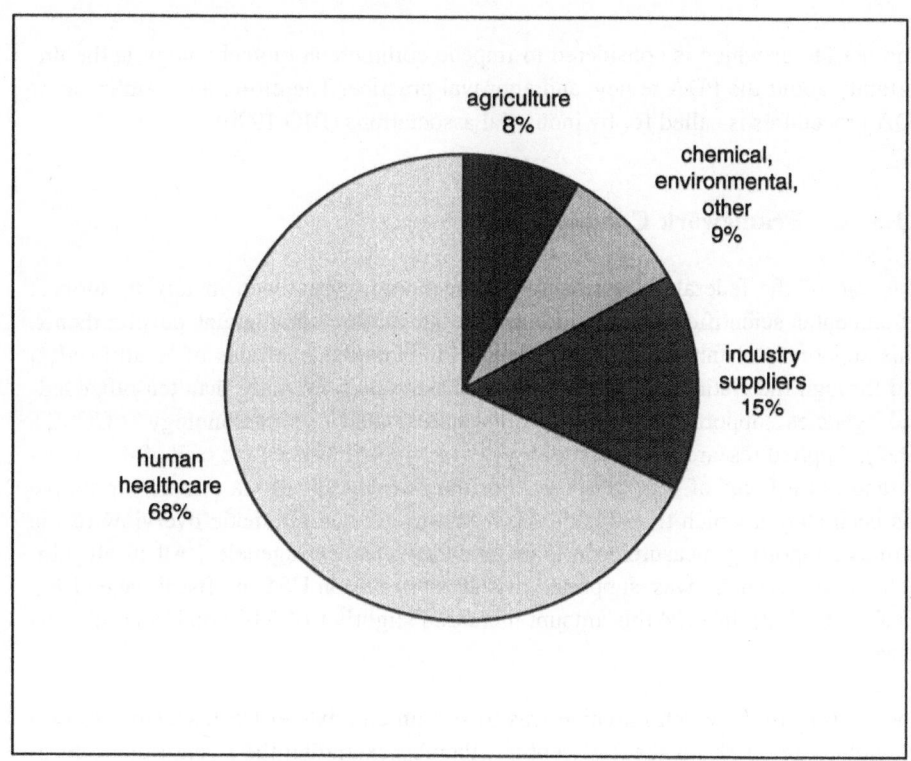

Most of the US biotech companies are small and medium-sized enterprises (SMEs). About 75 % of these companies have less than 50 employees. However, in contrast to most other countries there are also a few larger biotech enterprises which include some of the best-known firms in this business like Amgen or Genentech.

The focus of activities of American biotech firms is mainly the health care sector (BIO 1996). Almost 70 % of all biotech firms are active in this area (figure 2.1-1). About 15 % of the firms are suppliers for biotechnology. About 8 % are in the agro-sector and another 9 % are focusing on chemical, environmental and other applications of biotechnology.

The biggest problem for biotech firms in the United States at present is capital availability. For example, cash use for public biotechnology companies increased by 16 % in 1994 while at the same time cash sources increased by only 9 %. This has led to an almost 30 % decline in the survival index for the median public biotech company from 34 months to 25 months (BIO 1996). This means, for example, that 26 % of public biotech companies can expect to last less than one year at their current cash-burn rates. 50 % of these companies have only enough capital to last two years or less.

Another factor which is considered to impede commercial biotechnology is the uncertainty about the FDA review and approval practice. Therefore, a streamlining of FDA procedures is called for by industrial associations (BIO 1996).

2.1.3   Framework Conditions

The role of the federal government in biotechnology has been mainly to support fundamental scientific research as a basis for further commercial developments. This support is channelled mainly through the National Institutes of Health (NIH) and through the National Science Foundation. In addition, more than ten other federal agencies support to some extent research related to biotechnology (FCCSET 1992). Applied research and technology development has been the task of the industry and not a focus of federal policy. For the fiscal year 1993 a research initiative has been started which tried for the first time to give a systematic overview of the various supporting measures of the different government agencies. All in all, biotechnology research was supported by about 4 billion US$ in fiscal year 1993 (FCCSET 1992). In 1994 this amount increased slightly to 4.3 billion US$ (Gibbons 1994).

The goal of this research initiative was to sustain and extend US leadership in biotechnology research for the 21st century, thereby extending the scientific and technical foundations for the future development of biotechnology.

In addition to the federal support of fundamental research in biotechnology, other measures have been taken by the federal government (Gibbons 1994). One of those is the Advanced Technology Program (ATP) administered by the Department of Commerce which is designed to promote the economic growth and competitiveness of US businesses and industry by accelerating the development and commercialisation of promising high-risk technologies. The ATP is not specifically restricted to biotechnology. However, a considerable share of industry responses to ATP announcements come from the biotechnology area.

Federal laboratories have been instructed to devote a growing percentage of their budgets to R&D partnerships with civilian industry through, for example, cooperative, cost-shared research and development agreements (CRADAs). One of the first CRADA initiated by the NIH includes the NIH and Genetic Therapy Inc. This agreement is based on technology and expertise in the NIH laboratories in the field of gene therapy.

Finally, the government is also trying to improve private capital availability for biotechnology. For example, tax incentives for private sector investment in R&D and new business formation have been made law recently. Another effort includes the goal of making tax credit for research and experimentation permanent.

The use of genetic engineering in the United States is regulated by various regulations and recommendations made by different ministries and agencies (Hohmeyer et al. 1994). These are the Environmental Protection Agency (EPA), the Food and Drug Administration (FDA), the US Department of Agriculture (USDA), the National Institutes of Health (NIH) and the Occupational Safety and Health Administration (OSHA). There is no general genetic engineering act in the USA. The underlying principle is that the regulation of genetic engineering does not require specific laws and regulations but that existing laws and regulations can be adapted to the specific questions posed by genetic engineering. This fundamental rule was laid down in the co-ordinated framework for regulation of biotechnology in 1986. Another fundamental rule is that regulation has to consider the properties of a product or specific hazards rather than the fact alone that a product is prepared by using genetic engineering methods. This rule applies particularly to the area of diagnostic agents, pharmaceuticals and foodstuffs. It does not apply to the release of plants, animals or microorganisms modified by genetic engineering.

Concerning implementation of regulation in the USA it can be observed that relatively strict conditions and restrictions are imposed initially as long as only a small amount of experience is available for a particular application and its possible risks. However, these conditions are gradually relaxed as suitable and comprehensive investigations show with time that a risk which was not accurately specified in the first case has been overestimated.

As mentioned above, one of the key problems of American biotech companies is capital availability (Ernst & Young 1994a). In particular, during the year 1994 many biotech firms faced severe financing problems, related not only to the venture capital market but also to the stock market. Many venture capital companies retreated from biotechnology and concentrated on "more interesting" industries like information technology. In addition, the stock market became unfavourable for almost all biotech firms. These negative developments are not due to a particularly bad performance of the affected companies. On the contrary, 1994 can be considered as a rather successful year for biotechnology. For example, 17 new pharmaceuticals finished clinical trials and are ready now for market introduction (Ernst & Young 1994a). The unfavourable development of the capital market is mostly due to a nonspecific negative climate which is created for example through discussions on health care reform or through some failures of biopharmaceuticals during clinical developments which have been reported on intensively in the media.

2.1.4   Conclusions

The USA are clearly the leading nation in biotechnology, as indicated by several criteria (highest number of dedicated biotech firms, broadest and strongest research base, best patent position). This judgement is especially true for health care-related biotechnology. The role of the government in the past has been mainly twofold: support of basic research related to biotechnology and development of favourable framework conditions.

At least two key elements which contributed to the success of biotechnology in the USA can be identified. Firstly, there was a strong science push which could draw on a well-developed strong research base. Secondly, the technological and commercial potential of this science push was realised rather early by the private sector. Mainly two groups are important in this context: one group are the entrepreneurs who took the risk of exploiting the commercial potentials of biotechnology. The other group are private investors who risked investing private capital into this new business. So far for both groups their engagement paid off.

It is questionable, however, whether the American situation could be transferred to other countries. This holds true especially for the second part of the success story, the venture environment. Therefore, it might be necessary to develop other strategies for transferring science into commercial success. But it is also very clear that all these exploitation strategies cannot be successful if the first factor or prerequisite, a broad and strong science base leading to a permanent science push, is not available.

2.2   Japan

2.2.1   Introduction

At the beginning of the 1980s a great enthusiasm about the potential of biotechnology arose in Japan. The powerful Ministry of International Trade and Industry (MITI) declared 1981 as the year of biotechnology. Mainly large enterprises started biotechnology activities during that time (Schmid et al. 1995). At the beginning of the 1990s many of these large firms had to realise that the promises of biotechnology have not come true as fast as expected. As a consequence some of these companies, like for example construction companies, retreated from biotechnology. Recently, however, again a growing interest in biotechnology, especially in health care-related biotechnology, can be observed again in Japan.

2.2.2   Research Institutions

Biotechnology research in Japan outside industry is performed mainly in national research institutions, in universities and in research companies. The national research institutes of the different ministries play an important role in disseminating new projects and technology. They also participate in the start-up phases and provide training facilities for personnel from industry. A regional centre for national research institutes is TSUKUBA Scientific Town, a science city near Tokyo. In TSUKUBA the Science and Technology Agency (STA), the Ministry of International Trade and Industry (MITI) and the Ministry of Agriculture, Forestry and Fisheries (MAFF) are running most of their research institutes. Recently, science towns near Osaka and Nagoya have become more important.

The following national research institutes are doing research in biotechnology (Schmid et al. 1995):

- National Institute of Bioscience and Human Technology and National Institute for Advanced Interdisciplinary Science (MITI): focus on biochips, micro-machine technology, glycotechnology, marine biotechnology, genome analysis, biological CO_2 fixation and bioremediation.

- National Institute of Health, National Institute of Hygienic Sciences, National Cancer Centre, National Hospitals (Ministry of Health and Welfare (MHW)): focus on genome projects relating to gerontology, anti-cancer projects, gene therapy, AIDS prevention, food safety.

- Food Research Institute, National Institute of Agrobiological Resources, National Institute of Agroenvironmental Sciences (MAFF): focus on rice genome project, genome studies in agricultural animals, tissue cultures for plant and tree propagation, embryo transfer project, biosensors for the food industry.

All in all there are 97 national universities with 78 attached research institutes, and 411 private and public universities in Japan, which are overseen by the Ministry of Science and Culture (MONBUSHO) (Schmid et al. 1995). Biotechnology has become a popular research target in many of these universities. Research programmes concentrate for example on genome projects, cancer-related projects and environmental applications of biotechnology. A very important function of academia in biotechnology in Japan is its role in providing advice to the government through the countless committees under the science councils.

Table 2.2-1:   Japanese Research Companies Active in the Field of
                Biotechnology

Name of Company	Participating Industry	Main Target	Duration (years)	Guiding Ministry
Protein Engineering Research Institute, Osaka	13 companies	Establishment of protein engineering techniques	10	MITI
Research Institute for Environmental Technology for the Earth (RITE), Kyoto	60 companies	Global CO_2 reduction, bioreactor technology, biodegradable polymers etc.	10	MITI
Marine Biotechnology Institute, 3 locations	24 companies	Biochemicals of marine origin	10	MITI
Drug Delivery Systems Institute	7 companies	Targeted drug delivery systems	7	MHW
Institute for Advanced Skin Research	3 companies	Artificial skin systems	7	MHW
Biosensor Laboratories Co., Ltd.	4 companies	Optoelectronic sensors for in vivo monitoring	7	MHW
Seatechs Co.	3 companies	Aquaculture or large fish	5	MAFF
Turf Grass Co.	4 companies	Recombinant herbicide-resistant turf grass	7	MAFF

Source: Schmid et al. 1995

Research companies are organised by a guiding ministry and typically are co-financed by a consortium of at least three firms. They are typically founded for a life span of 7 to 10 years and are usually located in a building specially constructed for this purpose. Research companies relevant for biotechnology are summarised in table 2.2-1.

In the past most emphasis in Japanese research-related biotechnology has been on applied research and immediate commercialisation. This has led to a comparatively weak fundamental science base in biotechnology, which turned out to be an competitive disadvantage. However, there are also exceptions from this observation which have been put forward mainly through the STA. Among these especially the ERATO projects and the Frontier Research Programme deserve attention.

ERATO started as a programme in 1981 and supports a limited number of projects for an average period of five years, with an average budget of roughly 10 million US$. For each project a director is appointed who is responsible for the scientific content and management of the project. A typical research group is composed of 50 to 20 young scientists. The aim of ERATO projects is to do basic research in unexplored regions without any goal oriented restrictions. At present there are 18 ERATO projects running, about half of which are significant for biotechnology. These include for example questions of biofouling, cell growth and division, protein crystals, brain and behaviour, genome organisation, plant metabolites, bacterial motion systems, new extremophilic microorganisms.

In 1986 STA initiated the Frontier Research Programme. This is an internationally open programme and includes foreign project leaders. Presently the following research themes are covered: plant-homeostasis research, glycobiology research, new materials, brain mechanisms, and photodynamics.

2.2.3   Commercial Activities

The Japanese market for biotechnology is estimated at about 5.6 billion US$ for the year 1992 (JBA 1993). The health care sector makes about 3 billion US$ of this total volume. Chemicals amount to roughly 0.8 billion US$, plant production about 0.4 billion US$ and the food sector roughly 30 million US$. In particular, the health care market is expected to increase significantly during the coming years. These expectations arise for the following reasons:

- Japan is one of the biggest consumers of pharmaceutical products world-wide. The Japanese population makes up only about 2 % of the world population but accounts for about 20 % of the world's pharmaceuticals consumption.

- In 1992 the price system for pharmaceuticals has been changed. Prices are adjusted every two years, leading on average to a reduction of 8 %. This system fosters the development of new pharmaceuticals which can be placed at the market initially at a rather high price.

- The ageing Japanese population will have new health care needs which will lead to the creation of new markets.

Biotechnology in Japan is pursued mainly by large enterprises and trade companies which include biotechnology into their internal diversification programmes. The focus of these companies is on various fields like food production, brewing, tobacco manufacturing, pharmaceuticals or construction. In many cases these companies have been dependent on biotechnology know-how which has been generated in the United States. Strategic alliances with American biotech companies are a preferred way of getting access to this knowledge. Most of these partnerships have been marketing or licensing agreements (Bullock and Dibner 1994). Only about 18 % of the partnerships were focused on research and development (R&D). This indicates that the Japanese companies have been mainly interested in a rather rapid access to market by taking advantage of basic, already existing know-how. Building up own research capabilities and thereby generating generic know-how has been of minor importance. However, there seems to be a tendency towards an increasing significance of R&D partnerships (Bullock and Dibner 1994). This development could at least partly be caused by bad experience of Japanese large enterprises with American partners in the past. Accordingly, Japanese firms are more selective nowadays when choosing foreign partners.

Some Japanese companies choose a different way for getting access to American biotechnology know-how, namely by establishing research centres in the USA. However, these approaches have not been very successful so far. For example, Hitachi financed a research centre at the University of California providing space for 80 scientists. In 1994 only about 20 % of these positions had been filled. None of the employees is Japanese. In addition, the annual budget has been cut by 30 % (Valigra 1994).

The significance of small and medium-sized enterprises (SMEs) for biotechnology in Japan is discussed controversially. Following a rather broad definition of biotechnology, which includes traditional food production and fermentation processes, several hundred SMEs could be considered as biotechnology companies. However, if a more narrow definition of biotechnology is used, which is also preferred for example by the OECD or OTA, then only very few biotechnology SMEs exist in Japan. According to Bullock and Dibner (1994) there is only one Japanese firm which could be comparable to American biotechnology SMEs. In general, most experts agree on the notion that SMEs play no role in biotechnology in Japan.

One of the main reasons for this situation is that there is a generally conservative attitude within Japanese research centres about taking the risk of founding one's own enterprise. In other words, the entrepreneurship, which has been typical in the United States during the early stages of biotechnology, is not present in Japan. A second reason is the lack of venture capital for biotechnology. There are only very few venture capital firms in Japan and biotechnology plays almost no role within their portfolio (Kinoshita 1993). On the other hand, there is almost no demand for private venture financing. Biotechnology is supported mainly through public funding and more importantly, through large enterprises. The MITI has tried to improve this situation, for example by initiating a venture enterprise centre which gives loans to SMEs at very low interest rates. However, these and other approaches so far have had almost no impact on biotechnology. Due to the present problems in the Japanese economy it is expected that the readiness to start risky endeavours, like founding new biotechnology companies, will even decrease.

2.2.4   Framework Conditions

Japan has a long history as a centrally governed nation. Accordingly, the central government also plays an important role in shaping the economy, and in consequence also science and technology which are mainly seen as an infra-structural measure to enhance the economy (Schmid et al. 1995). There is, however, a complementation of this "top-down approach" through the influence of many councils where leaders from industry and research institutions formulate recommendations for government policies.

There is no comprehensive policy on biotechnology in Japan. On the contrary, at least five ministries and two governmental agencies are formulating their own policies for different areas of biotechnology. Quite often there is competition between the different initiatives. The formal co-ordination of science and technology policy in general and also in biotechnology lies with the Science and Technology Agency (STA) which is formally headed by the prime minister through the Council for Science and Technology. The STA has the biggest budget for biotechnology research and is for example responsible for the ERATO and Frontier Research Programmes. Other ministries competing with the STA include the MITI, the Ministry of Post and Telecommunication, the MAFF, the MHW and the MONBUSHO. The over-all budget related to biotechnology of these ministries and agencies amounted to Yen 120 billion in fiscal year 1993.

Genetic engineering in Japan is regulated by various regulations and recommendations made by different ministries and agencies. There is no special genetic engineering law in Japan at present (an attempt to regulate important areas of the application of genetic engineering which was made in 1991 by the Japanese environmental authorities failed, owing to the resistance of other ministries and agencies (Hohmeyer et al. 1994)).

Broad areas of Japanese regulatory policy are based on American models. Thus, the Ministry of Health and Social Affairs has to a large extent adopted the directives of the FDA in the area of pharmaceuticals produced by genetic engineering. The Ministry of Agriculture is also relatively strongly oriented towards the practices of the American Department of Agriculture. However, it appears in practice that the Japanese approval authorities proceed with greater caution regarding the possible negative effects of genetic engineering than is the case in the USA at present. For example, the first release experiments of transgenic plants were associated with demands for extremely thorough documentation of all conceivable effects, even though there had been already large international experience with such experiments.

The lack of a legal framework for genetic engineering in Japan does not mean that there are no binding rules. This is due to social conventions in Japan, which are different from western societies. Accordingly, directives and recommendations in Japan can entail a very high degree of social obligation. Hence a recommendation in Japan may have the same binding character as a law in western countries. The following ministries and agencies are involved in the regulation of genetic engineering:

- The MONBUSHO, which is responsible for the whole of government-financed research with the exception of research carried out in the institutes of other ministries. For example, all university research work falls within the area of responsibility of MONBUSHO.

- The Science and Technology Agency (STA), which is responsible for all genetic engineering research work in industry. However, production activities carried out in industry are not covered by the STA.

- The Ministry of Agriculture, Forestry and Fisheries (MAFF), which is responsible for the use of genetic engineering and also the release of transgenic organisms in the areas of agriculture, forestry and fisheries.

- The Ministry of Health and Social Affairs (MHW), which is responsible for the manufacture of pharmaceuticals and the production of foodstuffs.

- The Ministry of International Trade and Industry (MITI), which is responsible for the remaining areas of industrial production. At present this is almost exclusively concerned with enzymes and chemicals.

- The Environmental Protection Agency (EPA), which is also responsible for the release of all transgenic organisms. Whereas the responsibilities of all the other ministries and agencies already quoted are sharply separated from each other, the responsibility of the environmental protection authorities lies diagonally to the responsibilities of the other ministries and often acts as a second form of regulation.

At present there is no public debate about the use of genetic engineering in Japan. However, the consumer attitude in Japan can be considered as rather sensitive. Therefore it is an open question how the consumer will react in the future if, for example, genetically modified food will be introduced in Japan.

2.2.5   Conclusions

Biotechnology in Japan has been driven mainly by large enterprises and trade companies. Small and medium-sized companies have played no role. The strategy of the large enterprises in the past has been mainly to acquire basic knowledge generated in the United States and thereby enter the innovation system at a rather advanced stage. This strategy has proved to be successful in many other industrial sectors. However, in biotechnology so far this approach has failed. The Japanese experience in biotechnology points towards a more general conclusion. Research-intensive technologies with a strong science link require, on a firm level, the availability of own competitive research capacities, which provide the competent interface for communicating with research sites like universities where the needed basic knowledge is generated. Strategies which try to shortcut these linkages are not appropriate for biotechnology.

2.3 Germany

2.3.1 Introduction

Like most other industrialised nations Germany considers biotechnology to be a key technology for the 21st century. Accordingly there are various measures on the federal and on the state level for supporting the further development of biotechnology. Public support for biotechnology has a rather long history in Germany, starting in the early 1980s. In contrast to these public and also many private incentives in biotechnology the perception of biotechnology in Germany and abroad has been, and probably still is, rather critical. Mostly Germany has been considered as a location where it is very difficult to do biotechnology on an industrial level. A low public acceptance and impeding legal frameworks are often cited in this context. However, contrary to common belief, a broad variety of companies and research institutions which are active in different fields of biotechnology already exist in Germany today.

2.3.2 Research Institutions

There are many research institutions in Germany involved in basic science which to some extent is also significant for biotechnology. Such a broad understanding would mean that, for example, all university departments doing biological or medical research would be considered as biotechnological institutions. By such an approach the number of biotechnology institutions in Germany would be very high but this would also give a wrong impression of the biotechnology science linkage. The following estimates therefore are based on a more narrow understanding which excludes pure basic research without an application perspective. Under these preconditions there are presently about 550 research institutions in Germany doing research with high relevance to the industrial application of biotechnology (Statistisches Bundesamt 1995). About 80 % of these are university institutions, the remaining 20 % are Max Planck Institutes, federal research centres, state research centres, Fraunhofer Institutes and others.

Funding of biotechnological research at these institutions is rather complex. However, there is a common feature of almost all these institutions: contract research plays an important role. For example, in the state of Baden-Württemberg, the share of contract research in biotechnology at university is about 70 % of overall funding. The main share of contract research money comes from public bodies, however, roughly 25 % of these funds are already contracted by the industry (Reiss/Jaeckel 1994).

2.3.3   Commercial Activities

At present there are about 400 firms involved in biotechnology in Germany. Total sales figures of these companies are estimated at about 6.6 billion DM per year (ifo 1994). The structural analysis of these enterprises indicates that mainly two groups of companies can be differentiated (figure 2.3-1). Small and medium-sized companies with less than 100 employees and larger companies with more than 1000 employees.

Figure 2.3-1:   Structure of Biotechnology Companies in Germany
(Reiss and Hüsing 1992)

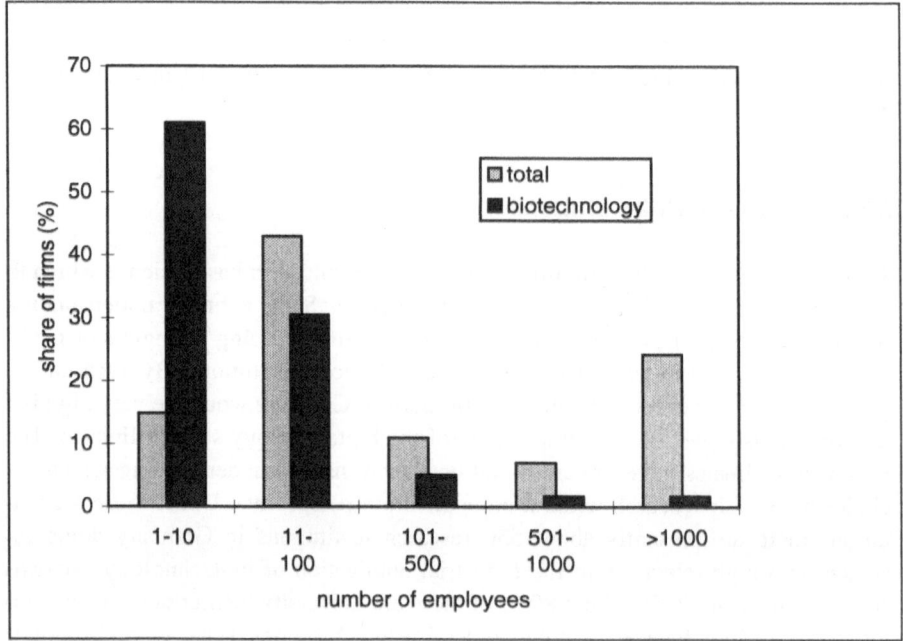

The share of this latter group is higher among biotech companies compared to the industrial average. However, if only employees who are working in the biotechnology units are considered, the size distribution is shifted towards lower numbers. This indicates that in larger enterprises biotechnology is being pursued in small units which are comparable in size with a stand-alone small or medium-sized enterprise. Looking at the company structure in terms of turnover, a similar structure is revealed. Again two groups of companies dominate: companies with less than 15 million DM yearly turnover and companies with more than 200 million DM turnover per year. Counting only turnover in biotechnology most companies make less than 15 million DM per year.

The age structure of German biotech companies indicates two groups as well. About half of the firms are more established and traditional firms who diversified into biotechnology. The remaining comprise new firms. Most of these new firms have been founded between 1985 and 1990.

Business activities of German biotech companies are directed mainly to the health care, to the environmental, to the agro-food and to the industry-supplies market segments (figure 2.3-2). In this respect there is a significant difference to US bio-tech companies, which concentrate mainly on the health care market.

Figure 2.3-2:   Market Segments of German Biotech Firms

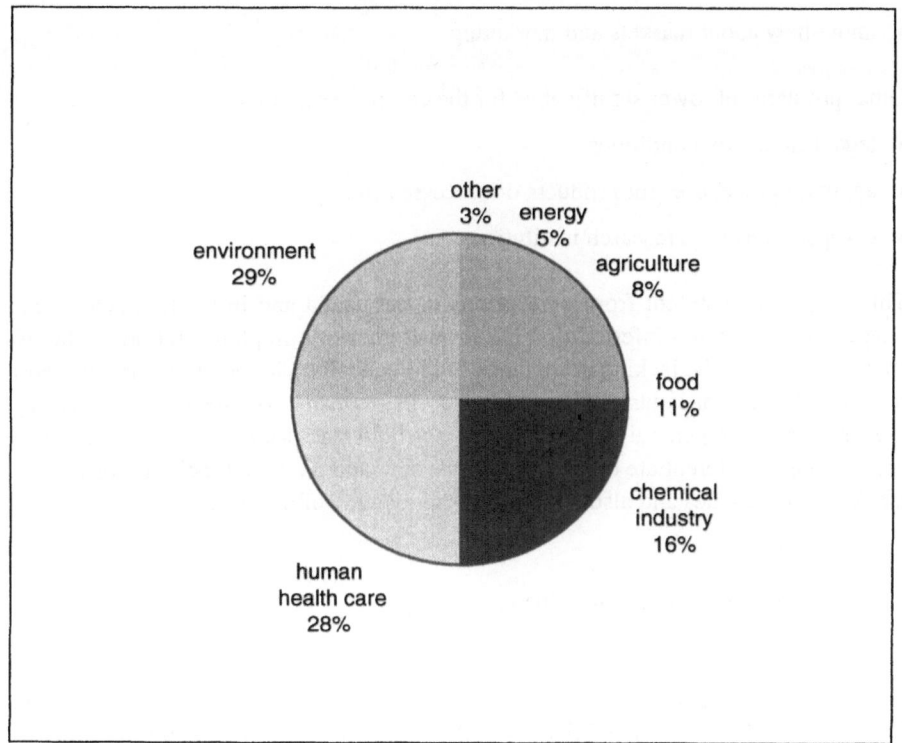

Research and development are crucial for biotech companies. German biotech companies invest between 15 and 17 % in R&D (Reiss and Hüsing 1992). With these figures biotechnology ranks as top of all industry sectors. For comparison, the average R&D intensity in German industry was 3.3 % and respectively 4.3 % in the chemical industry in 1989.

A key issue for industrial research activities in biotechnology are co-operations. About 80 % of all R&D projects are performed in some sort of co-operation with external partners. The most important co-operation linkages can be observed between companies and universities and also between different companies. This indicates that the innovation networks between knowledge-generating and knowledge-applying units are already well established in Germany.

According to a recent survey (FhG ISI 1995) small and medium-sized biotech companies are facing problems mainly in the following areas:

- R&D financing, which is especially difficult because of the very high R&D expenditure in biotechnology,

- market readiness of products which are being developed,

- financing of product marketing,

- know-how about markets and marketing.

Other problems of lower significance for the companies include:

- legal framework conditions,

- approval procedures for products or processes and

- co-operations with research institutions.

This ranking is different from perceptions in the media and in public discussions, where it is quite often inferred that the legal framework, implementation of the respective laws and the lacking acceptance of biotechnology in the public are the most impeding factors for industrial biotechnology in Germany. The results of the survey indicate that these general assessments do not hold true. Rather it is necessary and appropriate to differentiate these assessments depending on the type of company, on the focus of activities and also on the regional location of the firm.

2.3.4 Framework Conditions

Different federal and state agencies contribute to the funding of biotechnology in Germany. Most important of these is the Federal Ministry of Education, Science, Research and Technology (BMBF), which has developed a special programme for biotechnology: "Biotechnology 2000" which came into effect in 1990. Within this programme presently about 300 million DM per year are spent on biotechnology. Additional funds are available through different biomedical programmes. Recently genome research has drawn more attention within the public funding bodies. For example, the BMBF will support genome research by an additional 200 million DM. Other new approaches include the establishment of interdisciplinary clinical research centres at several locations. A new initiative of the BMBF supports the regional development of biotechnology. These federal activities are complemented through state initiatives in several of the German states. These include for example the support of new biotechnology products or innovation centres for start-up companies.

According to a recent survey by the Federal Statistical Office (Statistisches Bundesamt 1995) the overall expenditure for biotechnology in Germany is estimated to amount to about 2.7 billion DM. Public bodies and industry each contribute about 50 % of this total amount. In total, expenditure for biotechnology R&D accounts for about 3.5 % of the total R&D expenditure in Germany.

In contrast to public promotion measures for biotechnology, the private capital market in Germany has been weak. For example, in 1993 the overall volume of new venture capital for biotechnology in Germany amounted to only 23 million DM. This means that the share of biotechnology within all venture funding was only 1.9 %. By comparison, in the United States during the same year funds of about 1 billion DM have been spent on biotechnology, thereby allowing biotechnology to achieve a share of 25 % of all venture funding. This comparison indicates clearly that the limited availability of private financing is one of the most significant hindrances for biotechnology in Germany. However, just recently there have been signals that some private investors are developing a more favourable attitude towards biotechnology. For example, in summer 1995 a workshop on business opportunities for biotechnology for small and medium-sized enterprises was organised jointly by the FhG ISI and one of the big German private banks.

The most important legal framework for biotechnology in Germany is the gene law which implements the EC guidelines 219 and 220. The gene law was issued in 1990. In December 1993 the law was modified with the aim to facilitate and improve the legal procedures and at the same time keep the high safety standard. The legal requirements for genetic engineering works in closed systems depend on the safety category of the organisms and on the type of work being pursued. Almost 99 % of all respective activities in Germany belong to the lowest two safety categories, which at present pose no significant problems for the applicants with respect to the implementation. In general, there seems to be an agreement now in Germany that the gene law provides a workable basis for biotechnological activities. The widely discussed problems with the legal framework are more related to the interdependence with other legal requirements and also to different implementations within different states (FhG ISI 1995).

2.3.5   Conclusions

Compared to the United States, biotechnology in Germany adopted a different course. The German biotechnology industry comprises not only start-up firms, but also many established firms which diversified into biotechnology. In addition, the business focus of industry is more diverse: besides the health care sector, environmental applications, the agricultural sector, also the industry supply sector are covered by these companies. This different industrial structure has to be considered when comparing the German situation and competitiveness with the United States. It is not very meaningful to look just at the number of dedicated biotech firms, rather the whole picture has to be taken into account.

After doing so, the situation of biotechnology in Germany can be evaluated rather positively. There is a broad and strong R&D infrastructure, personnel is highly qualified, the German and European target markets are attractive. The legal framework is now considered mostly as acceptable, public funding is available, not only from federal, but also from state and European sources. Main problems are private financing and the speed and efficiency of know-how and technology transfer. Considering the strong relation between science and commerce in biotechnology it is not only important for the future to improve the availability of private capital for biotechnology and the transfer of know-how from science to industry, but also to maintain a broad and competitive research base. In general, the whole process of innovation needs to be checked for bottle-necks which then in turn give clues for further optimisation.

2.4   United Kingdom

2.4.1   Introduction

The United Kingdom is considered one of the most favourable locations for biotechnology within Europe (Ernst & Young 1994b). For example, the UK has the highest number of biotechnology start-ups in Europe and is the most favoured European investment location by executives outside their home market. One important reason for this role of the UK is its traditional high quality of life sciences (OTA 1991). Some of the most important methods and technologies for genetic engineering have been developed in research institutions in the UK. Among these are, for example, sequencing techniques for DNA and proteins and also the basic procedures for monoclonal anti-body production. Another factor favouring biotechnology in the UK is the financing environment. The UK has a very active venture-capital community. For example, in 1993 venture capitalists in the UK were responsible for more than half the money provided for biotech investments in Europe (Ernst & Young 1994b). In addition, changes to the London stock exchange listing rules have made it easier for biotech companies to get on the main list thereby providing additional funding opportunities and also an exit route for venture capitalists.

However, there have also been dramatic ups and downs of commercial biotech activities in the UK. For example, after the enthusiastic situation at the beginning of 1993 in the context of the better accessibility of the London stock exchange, UK biotech stock prices have fallen dramatically during 1994 and also initial public offerings have been postponed widely (Spencer and Kirk 1994).

Another interesting feature of biotechnology in the UK is the public debate. Namely, in 1994 the UK followed the examples of Denmark and the Netherlands in organising a consensus conference on the subject of plant biotechnology. This conference aimed to increase public debate about applied science and to explore public attitudes to applications of plant biotechnology. During this exercise scientists and the general public were brought together in an auditorium where both points of view could be discussed openly and honestly. As a result of this conference, a lay panel of 16 members formulated recommendations for the field of plant biotechnology which may serve as a basis for the future communication of biotechnology between the public and the involved actors.

2.4.2 Research Institutions

Biotechnology research in the United Kingdom is performed in many universities, colleges, public and ex-government research centres as well as agricultural research stations which cover a broad spectrum of basic and applied activities. According to the UK Biotechnology Handbook '95 (BioCommerce Data Ltd. 1995), there are about 110 academic institutions and 28 UK government departments and laboratories active in biotechnology research.

Since the 1980s the funding system and accordingly the framework for co-operation between public research institutions and industry has changed a lot. The most recent shift towards a closer relationship of science and industrial applications occurred since 1992 when the White Paper "Realising our Potential" was published. Institutional funding was cut back further and the funding bodies for higher education of the regional education departments were reorganised. Universities now get tuition fees for full-time and EU students from the government and teaching grants from the higher education funding council. General funding for permanent academic staff and research facilities costs is also provided by the education departments and allocated by the regional higher education councils. A considerable share of the council's budget has been transferred recently to research councils which led to a potential overall loss of funds for universities (Dickson 1995a).

In general, government funding of university research is shared by the education departments and the Office of Science and Technology (OST) (dual support system). The education departments provide general funding for research while the OST is responsible for specific funds which are channelled through six research councils. The OST and the six research councils have been transferred from the Cabinet Office to the Department of Trade and Industry in 1995 with a Minister for

Science and Technology responsible to the president of the Board of Trade (CVCP 1995).

R&D in biotechnology is funded mainly through two of the six councils: the **Biotechnology and Biological Sciences Council (BBSRC)** and the Medical Research Council (MRC). The BBSRC was established on 1st of April 1994 by the incorporation of the former Agricultural and Food Research Council with biotechnology and biological science programmes of the former Science and Engineering Research Council. Like all research councils the BBSRC has its own mission:

- to promote and support high quality basic, strategic and applied research and related post-graduate training relating to the understanding and exploitation of biological systems,

- to advance knowledge and technology, and provide trained scientists and engineers, which meet the needs of users and beneficiaries, thereby contributing to the economic competitiveness of the United Kingdom and the quality of life,

- to provide advice, disseminate knowledge, and promote public understanding in the fields of biotechnology and the biological sciences (BBSRC 1995).

In 1995/96 the budget of the BBSRC amounts to 172,3 million £. In real terms this is about the same figure as in the previous period. The distribution of the budget by scientific areas (as in April 1995) is shown in figure 2.4-1. (BBSRC 1995). About 60 % of the budget is allocated to science-led programmes which cover six major areas: biomolecular sciences, genes and developmental biology, biochemistry and cell biology, plant and microbial sciences, animal sciences and psychology, engineering and physical sciences. The remaining 40 % of the budget support three directorates with the specific purpose of promoting strategic programmes of high-quality science to underpin the needs of the user communities: Agricultural Systems Directorate, Chemicals and Pharmaceuticals Directorate, Food Directorate. The directorate programmes provide a bridge between fundamental science and the applied needs of users in industry and government.

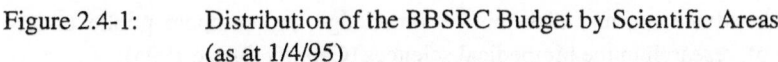

Figure 2.4-1:   Distribution of the BBSRC Budget by Scientific Areas
(as at 1/4/95)

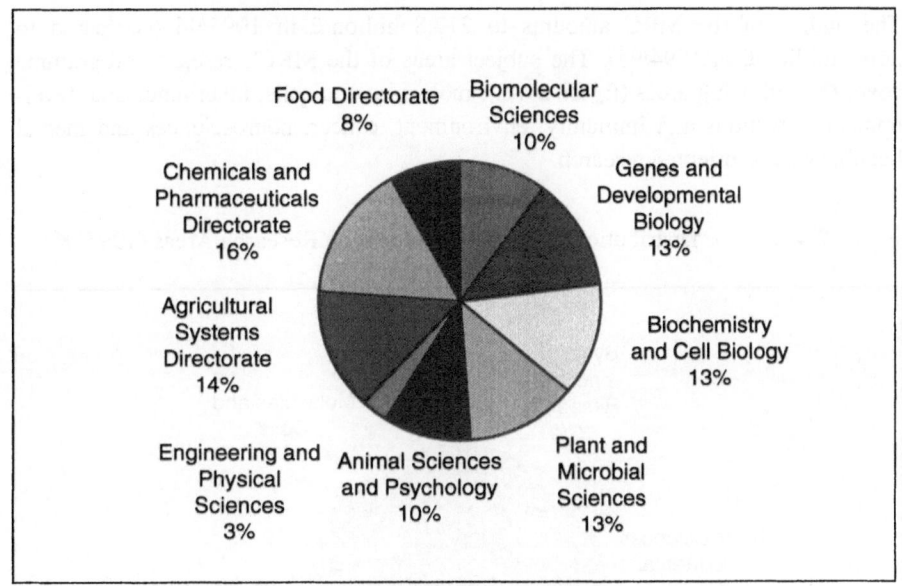

Besides its responsibility for supporting the science base in universities, other research centres and institutes, the BBSRC has complementary responsibility to spon-
sor eight agricultural, food and biotechnology research institutes. These institutes
are independent companies with charitable status. They are considered as national
centres of expertise and combine longer term strategic research programmes with
more specific topics, the latter often funded by industry and commerce. On average,
the BBSRC funds roughly 50 % of the institutes' budgets. The BBSRC sponsored
institutes are: Babraham Institute, Institute for Animal Health, Institute for Arable
Crops Research, Institute of Food Research, Institute of Grasland and Environmental Research, John Innes Centre, Roslin Institute, Silsoe Research Institute.

In addition to these institutes there are also four interdisciplinary research centres
(IRC), which are sponsored by BBSRC: Advanced Centre for Biochemical Engineering, Centre for Genome Research, Oxford Centre for Molecular Sciences, Sussex Centre for Neuroscience.

The mission of the **Medical Research Council (MRC)** is: to promote and support
high-quality basic, strategic and applied research and related post-graduate training
in the biomedical and other sciences, with the aim

- of maintaining and improving human health,

- to advance knowledge and technology and provide trained researchers which
  meet the needs of users and beneficiaries,

104

- to provide advice on, and disseminate knowledge and promote public understanding of, research in the biomedical sciences (Cabinet Office 1995).

The budget of the MRC amounts to 277,8 million £ in 1995/96 (compared to 266.7 million £ in 1994/95). The subject areas of the MRC's research programme cover the following areas (figure 2.4-2): molecules and cells, inheritance and development, infections and immunity, environment, cancer, neurosciences and mental health, systems oriented research.

Figure 2.4-2:   Distribution of the MRC Budget by Research Areas (1995/96)

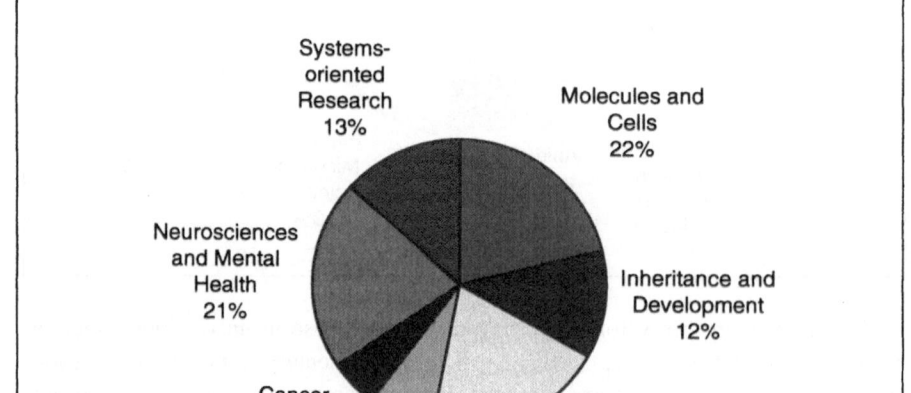

The MRC is also playing an increasing role in technology transfer. In 1993/94 the total income from the council's research collaboration with industry was about 7 million £. 29 new patent applications were filed in 1993/94. 139 licence agreements were enforced by the end of March 1994. The council has also increased its activities in working with sources of new capital to extend its role in the creation and development of new companies. Examples for these activities include Cambridge Antibody Technology Ltd. and Therexsys Ltd. (Cabinet Office 1995).

Research activities of the council are concentrated in three large institutes (the National Institute for Medical Research, the Laboratory of Molecular Biology and the New Clinical Sciences Centre), where the council employs its own research staff. Most of the research groups within these centres are embedded within universities. Recently, a new research centre, the Edward Jenner Institute for Vaccine Research, has been established, which is a jointly funded centre supported by the MRC, the BBSRC, the Department of Health, Glaxo and the Welcome Trust.

2.4.3   Commercial Activities

The British biotechnology industry is considered one of the best developed in Europe. However, as is the case for most other countries, consistent and comparable data about the British biotech industry is hardly available. The following information is compiled from different sources which use different definitions and understandings of biotechnology.

According to the most recent edition of the UK Biotechnology Handbook (BioCommerce Data Ltd. 1995) biotechnology dependent sales by UK user industries and biotechnology SMEs together account for over 4 billion £ in 1995. It is predicted that sales by UK industry will increase to over 9 billion £ by the year 2000. Biotech start-ups have contributed about 400 million £ in 1992 to these over-all sales in biotechnology and are expected to increase this figure to about 1 billion £ of sales by the year 1996 (Ward 1994a). The total workforce of the biotechnology sector was 7,400 in 1993 and is estimated to increase to about 11,000 by 1996 (Ward 1994b).

At present the number of biotech companies in Britain is estimated at about 380 (Biocommerce Data Ltd. 1995). This number includes companies which are using biotechnology directly or producing products used by the industry, such as equipment or reagents. In addition, there are about 175 companies providing services such as consultancy, legal advice, patenting services, culture collections and so on. Considering a survey of biotechnology start-ups in 1992 which estimated the number of start-up companies in biotechnology at about 170 (Ward 1994a) it can be concluded that about half of the current biotechnology companies in Britain are start-ups. Among these are some of Europe's largest and most advanced companies. For example, British Biotech in terms of market capitalisation is Europe's largest biotech start-up (Ernst & Young 1994b). In 1995 British Biotech had five potential pharmaceutical products in human clinical testing. Two of these have already reached phase II and III. As of March 1995 British Biotech employed about 300 people. In addition to these start-up companies there is also a number of large multinationals in the UK with substantial biotechnology interests. These include Zeneca, Glaxo, Smith Kline Beecham, Welcome and Unilever.

However, the overall profile of British biotechnology companies is characterised by a majority of small and medium-sized firms. About 50 % of these firms employ less than 50 people, about 80 % less than 100 (Dickson 1993). Most British biotechnology start-ups are mainly in the pharmaceutical sector. About 23 % of these companies are developing drugs and vaccines, about 25 % are concentrating on diagnostics. Another focus is on agricultural biotechnology, where about 13 % of the start-up companies are active (Ward 1994a).

2.4.4   Framework Conditions

As outlined above (see chapter 2.4.2), there have been recent shifts in biotechnology in Great Britain towards strengthening the relationship between science and industrial applications. This move is reflected, for example, in the establishment of the three strategic programme oriented directorates of the BBSRC. In addition, the UK government has established in its Department of Trade and Industry a task force for biotechnology, to take care of biotech companies (Ernst & Young 1994b). Another initiative aims at improving the awareness of industrial sectors of the potentials biotechnology might hold for them. According to a recent review commissioned by the Department of Trade and Industry, much of industry is unaware of how biotechnology could improve production processes, waste management and product development (Biocommerce Data Ltd. 1995). To address this lack of awareness a new DTI initiative was launched in spring 1995: "Biotechnology means Business" (BMB). The initiative aims to inform companies about biotechnology, establish contacts between these companies and suppliers and experts who can help them to explore biotechnology as an additional means within their company's strategy. So the over-all strategy of the BMB initiative is to function as a broker between user and supplier.

Other programmes aiming at improving collaborations between industry and research are the link scheme and the collaboration with industry scheme (CWIS), which fund projects in several biotechnology areas. These programmes are usually jointly funded by different government departments and the biotechnology-related research councils.

A new impetus for promoting biotechnology-related research activities in Britain may come from the recently started foresight activities. The results of the British Delphi survey have been discussed by 15 sectoral expert panels. They formulated 316 detailed recommendations, out of which six cross-sectoral strategic priorities have been identified. One of these priorities is the developing of new organisms, products and processes from genetics. In addition, genetic and biomolecular engineering and bioinformatics are considered as some of the most important generic priorities in science and technology (Dickson 1995b).

The government approach towards the legal framework for biotechnology is characterised by a deregulation policy. For example, the UK has been among the first to streamline the requirements for field testing genetically modified crops, has been among the fastest in Europe to approve genetically modified foods and also at the

EU level the UK has argued, for example, against the marketing moratorium on Bovine Somatotropin (BST) (Ernst & Young 1994b).

Private financing for biotechnology in the UK is considered to be better developed than in other European countries, especially the UK venture capital community is mentioned in this context. Also, the opening of the London stock exchange for biotech companies is considered an important improvement of private financing (Ernst & Young 1994b). However, according to company representatives (Ernst & Young 1994b), it is still very difficult to raise finance in the UK for a company at its very early stages. Investors in the UK are considered not as venturous by far as in the United States. Other problems which are seen for the future of biotechnology in Great Britain include the brain drain to the United States. Already today a rather high proportion of US biotech companies are led by management from the UK.

2.4.5   Conclusions

Biotechnology in the United Kingdom combines elements which can also be observed in other European countries plus typical features of the American situation. For example, the industrial structure in biotechnology, as in Germany, comprises one half of mostly start-up companies and the other half more established companies which have diversified into biotechnology. Political measures as in other European countries try to improve technology transfer and also increase the awareness of other industries for the potentials of biotechnology. The research base in life science in Great Britain is very strong, as in the United States and also some other European countries. Private financing in Great Britain is considered the best developed among the European countries. This situation, however, has also led to similar problems as in the USA: the stock market, for example, is reacting very sensitively to success or failures of clinical drug developments. Another problem for the future of biotechnology in Great Britain is seen in an increasing brain drain to the United States which is also facilitated by strong similarities in the two countries.

2.5   Israel

In 1993 about 30 biotechnology firms existed in Israel (Frenkel et al. 1994). Most of them are small companies which are based on products or processes developed in research, done in academic institutions. Most of these firms were set up as subsidiaries of research institutes or universities, some others are subsidiaries of foreign companies. Only a minority has been established with venture capital. Most of the companies are based on technological knowledge and skills of a single academic researcher.

Since then the biotech business in Israel has been growing considerably. In 1996 there are about 150 biotechnology firms in Israel. Their focus is on pharmaceuticals and medical products, human diagnostics, cell and gene therapy, veterinary products, agricultural products, marine and aquaculture, environmental biotechnology and research products (Brower 1996). Only 10 % of these companies are larger companies with over 100 employees. 24 % are medium-sized with 20 to 99 employees and the remaining are small firms with less than 20 employees (Brower 1996).

The following data describe in more detail the diagnostic companies which have been analysed recently during an empirical study (Frenkel et al. 1994). Most of these companies were established after 1980. All have by 1990 succeeded in producing and marketing their own products. All of these companies are small with an average number of 23 employees. R&D expenditure is rather high, about one third of total spending go to R&D. These companies rely heavily on exports. About 75 % of all sales are exported. The main market is Europe, with two thirds of the exports going to that market. The United States is not a principle market, except for those companies that are wholly owned subsidiaries of American companies.

The following factors are considered to pose constraints to the development of biotechnology in Israel:

- lack of venture capital for establishing new firms,
- lack of venture capital and other forms of risk capital for existing companies,
- lack of academic research centres specialising in biotechnology,
- lack of trained manpower in biochemical engineering, production and management engineering,
- lack of technological infrastructure in existing drug and chemical companies that use traditional technology,
- the small size of the local market in Israel coupled with the large distance from foreign markets (Frenkel et al. 1994).

In order to remedy some of these restrictions, a national biotechnology programme has been established.

From the scientific and technical point of view the biotechnological industry in Israel is considered to be advanced and competitive. However, many of these companies are facing problems in the transition from the R&D stage to the market. These problems are mainly due to a lack of capital and difficulties during market access. This situation could finally pose a danger that Israeli expertise in biotechnology will be recognised by foreign firms, who will then purchase it, bringing the employment, exports and profits to a company outside of Israel. On the other hand, a growing recent interest of US firms in Israeli biotechnology companies has led to the formation of strategic alliances and mobilisation of private capital for the Israeli biotech industry (Brower 1996).

3.   State and Development of Hungarian Biotechnology

This chapter describes the core part of the Biotechnology Audit the analysis of strengths and weaknesses of this technology field as well as framework conditions, e.g. legal framework, intellectual property rights. It includes such areas as university/industry linkages, and touches on public acceptance.

In the beginning of this chapter the legacy of the socialist system in the field of biotechnology is described in order to provide background information for the understanding of the present situation. The second section analyses the framework conditions for biotechnology in Hungary. The third section describes the present situation on the basis of the empirical study.

3.1   Legacy of the Socialist System in the Field of Biotechnology

For the 45 years prior to 1989, Hungary's economic structures were those of the communist system, because of the political and economic domination of the former Soviet Union. As it restructured its economy in the 1950s, Hungary, like other East European socialist countries, followed the Soviet model of central planning and management based on Marxist-Leninist theory and the economic laws of socialism. In the late 1980s Hungary was engaged in radical economic, political and social change. Since 1968 she has made numerous attempts to overcome the main feature of the socialist system: the lack of innovativeness since economic actors had little interest in commercialising R&D results. In the field of S&T policy formulation, the most important attempt was to re-link R&D and business activities, in order to shorten the path from inventions to practical application of new technologies in the production process.

This rethinking process resulted in the Large-scale Development Programmes that were launched from the 1970s. They represented a modified bureaucratic approach to research programmes and they were able to encourage a shift away from the interrupted innovation model to the one-way linear innovation model. The reformed socialist model was able to achieve that different stages of innovation should be considered in a dynamic context and not separately. But this transformation was not enough to approach the feedback loop model of innovation (Freeman 1982, Kline/ Rosenberg 1986). It also moved from autarchy towards international collaboration, but participation in international networks in S&T was restricted. Nevertheless, this reform helped to establish direct horizontal linkages and direct co-operation between institutes, universities and business, slowly replacing vertically organised bureaucratic connections. Business enterprises and production areas were less isolated from the domestic R&D sector, and the gap between operation of technology used and state-of-the-art technology diminished somewhat. But incentives for commercialisation and innovation remained weak, hence development and testing activities were inadequate.

Biotechnology was one of the targeted groups of Large-scale Governmental S&T Programmes. In the case of biotechnology it is important to follow an approach which is oriented towards the whole innovation process. Public support for biotechnology research has a rather long history in Hungary. In 1982 the Science Policy Committee requested the following ministries and governmental agencies to prepare a biotechnology R&D programme:

- Hungarian Academy of Sciences (HAS),
- National Committee for Technological Development
- Ministry of Agriculture and the Food Industry
- Ministry of Industry
- Ministry of Health

During the programme preparation phase their task was to analyse the situation of research and application in the field of biotechnology in Hungary. Then they had to prepare recommendations for a biotechnology mission taking into account the requirements of the planned economy. The Science Policy Committee approved the programme for "Biotechnology Research, Development and Application in Agriculture and Industry". The programme started in 1984. It tried to take into account multidisciplinarity and the importance of the university/industry link of biotechnology. The key features of the programme are shown in table 3.1-1.

The Hungarian biotechnology programme for the Third World will not be discussed in the following. However, it is worth mentioning that Hungary has a potential in the fact that exploitation for developing countries is supported by the UN (training centre etc.).

Commercialisation of biotechnology was included in the first government programmes. But it is very difficult to follow the impact of the programmes because the transition period greatly affected the organisations that were involved in it.

Table 3.1-1: Top and Down Governmental Programmes for Biotechnology
(before Transition Period 1984-1990)

Framework:	- OKKFT (Public Medium-term Research and Development Programmes during 6th five-year plan period) OKKFT A/16 - GF (Public Medium-term Research and Development Programmes during 7th five-year plan period) - G3 "Protein Programme"
Responsible government agency:	- Protein and Biotechnology Technical and Economic Advisory Board of Biotechnology at OMFB
New research organisations set up:	- Agricultural Biotechnology Centre Gödöllő (founded by Ministry of Agriculture in 1986, scientific work started in 1990)
Education programme:	- Start to educate engineers for biotechnology at Budapest Technical University, Agricultural University Gödöllő; - Support education of up-to-date biotechnology at universities in natural sciences (started Budapest, Debrecen, Szeged), - Agricultural sciences (started Gödöllő and Pannon University, Keszthely, Horticulture and Food Processing University), - Animal science university
Knowledge distribution:	- Education programmes were supported by a special training package, video programmes, slides launch publications - Series: Folio Biotechnological - Biotechnology in our days - Annual issue in English of Hungarian articles published by Hungarian journals
International relations:	- Bi-lateral agreement: Austria, Finland, US, Czechoslovakia, GDR - Membership: ICGEB (International Centre for Genetic Engineering and Biotechnology in 1983 - Direct linkages with many foreign research institutes and universities

Sources: OMFB reports

The key targeted sector of the biotechnology programme was agriculture. Commercialisation in this area was expected to be handled by large state-owned farms, and large and medium-sized agricultural co-operatives. In the pharmaceutical industry, the programme was able to achieve very few remarkable results. Even if the scientific results were good, companies could not solve scale up problems.

By the end of the programme some remarkable results were obtained in the scientific and business fields. But some valuable R&D results were never commercialised. The reasons for non-commercialisation are diverse, for example, large cooperatives engaged in plant cultivation disappeared because of the transition and there were no longer any suitable business organisations for commercialisation. This issue will be discussed later in more detail.

The first top-down biotechnology programme was finished by 1990, coinciding with the beginning of the transition period. The public thinking on S&T and innovation policy changed substantially. Top-down policy formulation was no longer an issue (see in OECD 1993, Inzelt 1995c).

3.2 Framework Conditions for Biotechnology in Hungary

Although changes have taken place more satisfactorily in Hungary than in most Central and Eastern European countries, it is still proving to be a difficult and painful process. Hungary is moving away *from a planned economy and towards an open market-oriented economy*. In the following a brief overview of the transition process will be given (Inzelt 1995a).

"Since Hungary had already embarked on a wide range of reforms during the 1980s, (notably banking reforms, tax reforms, and the beginning of price and trade liberalisation), the Government elected in 1990 chose a gradualist approach. However, this does not imply that the reform process was slow. Quite the contrary, as witnessed by the pace and rigour of structural policy." (OECD 1995a, p. 12).

Hungary has a multi-party system, with a *unicameral* Parliament. The governing coalition of the Hungarian Socialist Party (MSZP) and the Alliance of Free Democrats (SZDSZ) is proceeding with a solid and ambitious economic austerity programme since 1994. The basic aim of the freely elected governments and Parliament (1990, 1994) has been to continue the market-type reforms already undertaken and to accelerate the transition to a market economy. With 72 % of the seats in Parliament, it has a large majority to pass any legislation, including amendments to the Constitution. The governmental structure was redrawn to adjust to a market economy. The ministries were merged and new governmental agencies were established. The Planning Office, established in 1949, was incorporated into the Finance Ministry in 1990. The Ministry of International Economic Relations, established at the beginning of the transition period, merged with the Ministry of Industry and Trade in 1994 following the election of the new government. The structure is in an unsettled form, which reflects both deregulation and re-regulation tasks.

An important aspect of the new government is legislation. In establishing the new legal framework, Parliament has had to function like a "legislation factory." Numerous laws have been enacted since 1990, but this legislation process is not yet finished. For the purposes of this study, the following laws were important:[12]

1990

- Securities act
- Law on Privatisation (A new privatisation law is currently before Parliament. The objectives are to speed up the process.)
- Law on Prohibition of Unfair Market Practices
- Law on Protection of State Assets, Establishment of the State Property Agency
- Act on State Audit Office
- Stock Exchange Law

1991

- Central Bank Law
- Banking Law
- Law on Accounting
- Compensation Act
- Food Act (the Food Code)

1992

- Establishment of State Holding Company
- Labour Act

1994

- Law on Chambers
- Law on Hungarian Academy of Sciences
- Law on Higher Education

The function of these laws is to rebuild the fiscal system, reorganise and restructure corporations, and to continue and expand the market reforms of existing programmes, such as privatisation, trade policy and labour policy.

---

[12] Several important legal changes were initiated during the 1980s: a law on bankruptcy (1986, amended in 1991, 1993) and a two-tiered banking system (1987); personal income and value-added tax systems (1988); act on economic associations (1988); law on foreign direct investment (1973, 1989 amended in 1990); and the transformation act (1989).

Since the first free election (1990), Hungarian society has sought to break with old policies and establish new ones. The process of learning to formulate market-type strategies caused a gap in industrial policy. The economic policy concepts of the Hungarian government have changed radically many times in recent years. In 1993 the government announced its industrial policy and asked the OECD to review Hungary's industry and industrial policy. The view of the latter reflects modern thinking on the issue, which means that the review contains mainly *policies for industry*, including innovation and regional policies. The regional component of industrial policy is strengthening. (OECD 1995b) Furthermore, the government approved an Innovation Policy focusing on a few strategic directions: cost-saving and resource-saving technologies, biotechnology, electronics and information technology, and bio-active materials.

So biotechnology was among the small number of scientific and economic fields that were declared priority fields for Hungary, but it also became a part of the new S&T bidding system, bottom-up policy. In the case of bottom-up policy, programme initiators were microeconomic actors as opposed to the earlier top-down policy where initiators were macroeconomic actors. The programme itself has a rather complex structure. It covers basic and applied research, experimental development, education programmes, knowledge distribution, and international relations. Table 3.2-1 shows the importance of biotechnology among government-supported projects.

Table 3.2-1:   Bottom-up Governmental Programmes for Biotechnology
                 (After Transition Period, since 1990)

Area	Activity
Applied research, experimental development and early commercialisation:	- Framework: OMFB bidding system since 1991; - Programme: no special programme, but biotechnology is a favoured direction; - Governmental agency: OMFB (an S&T funding agency)
Basic research:	- Governmental and non-government financed research organisations and OTKA (public fund) grants; - International R&D co-operation in biotechnology government support 200 mio. HUF/year; - Target: commercialisation of biotechnology in 22 fields; - Number of contracts: 292

In summary, the results of the ongoing transformation have shown that Hungary is gradually moving towards a market-type policy formulation. However, policy-makers have concentrated more on the conditions of strategy than on the formulation of strategy. The learning process of formulating market-type strategies caused a gap in S&T (science and technology) and industrial policy (Inzelt 1996a).

3.2.1   Financing Issue

The availability of the critical mass of financial sources on time is a great challenge for actors in biotechnology (see chapters 2 and 4). The commercialisation characteristics of biotechnology are encouraging a number of changes in financing. Various financial instruments to raise the capital needed are explored. Internationally, going public is a source of financial means not only at the time of flotation of biotechnology companies.[13]

It is not surprising that these diversified financial instruments are not yet present in the Hungarian financial market which is still being built up. Hungarian monetary policy during the transition is caught between the dilemma of macroeconomic stabilisation and growth promotion while the capital market is in its infancy. Financial institutions do not show fully market-based behaviour (interest rates are very high, institutional investors such as pension fund are missing, commercial banks have not consolidated yet; limited market entry for foreign institutions, indebtedness, insolvency, loan losses etc.). Summing up, the financial market cannot offer favourable conditions to firms devoted to biotechnology.

Since the beginning of the transition period, business R&D expenditure have diminished rapidly (Inzelt 1995b). Generally, business is reluctant to purchase R&D. Besides the reasons mentioned above, the privatisation process, restructuring of the agricultural organisations, and health care reform also have negative impacts on financing biotechnology R&D activities by business, at least in the short and medium term.

---

13 For example, the London Stock Exchange (LSE) has adopted new rules for biotech (see chapter 2). They make it easier for biotech start-ups to sell shares to the public, and new markets. An Alternative Investment Market (AIM) was set up for small and growing businesses. It gives access to capital markets to early stage companies, which have little chance of meeting the full listing requirements of the LSE. It also aims to offer an earlier exit route to venture capitalists. The financing demand of biotech business also strongly encourages the establishment of a secondary market, a European version of NASDAQ. Venture capital financing is also important for this sector. As Ernst & Young (1994b) stated "European venture capitalists are keeping an eye on the sector. European CEOs agree that restricted access to capital has been a major inhibitor of biotech start-up formation. Many European venture capitalists have preferred to invest in US biotech firms because NASDAQ provided a visible exit route for them to realise profits from their original stakes." Different forms of venture capital organisations related to biotechnology are also an important element.

In such a situation the government can be the key financier of R&D. In this part of the study there will be no evaluation of the role of government. The focus is on the present financial conditions which are provided by the Hungarian government.

In real terms, public and business expenditure on R&D are declining. As the decrease has been faster in business financing, the state proportion has increased in relative terms (table 3.2-2). Other financial sources are very limited. Venture capital type organisations have appeared in Hungary but they usually do not engage in financing new high tech organisations. According to performance the relative proportion of higher education has almost doubled while the relative share of the business sector decreased considerably.

Governmental funding of biotechnology is only reported for one governmental fund, KMÜFA (Central Technical Development Fund). OTKA (Hungarian Scientific Research Foundation) gives many small grants to research organisations, however, its documentation is not available by field of science. Table 3.2-3 shows the number of contracts and the funding support by government in the framework of the new bidding system. The OMFB bidding system provides 50 % of project expenditure of firms and institutes on the condition that we have to pay back this support (in real terms, it is an interest-free credit).

Table 3.2-2:   Breakdown of Funding and Performance of GERD
(1988-93, in Percent)

	Funding[1]				Performance[1]		
	Business	Government	Private non-profit[2]	Foreign	Business	Government	Higher education
1988	76.6	21.1		0.9	51.4	15.9	11.3
1989	72.5	25.3		0.7	44.1	18.2	12.6
1990	70.1	28.9		1.0	38.1	19.5	14.4
1991	56.0	40.0	0.1	1.8	41.4	24.5	20.3
1992	52.5	41.5	0.2	2.9	36.5	25.3	21.4
1993							

Source: Central Statistical Office

Notes: [1] Calculated according to OECD institutional classification on the basis of

UNESCO one: business sector = productive sector (integrated and non-integrated R&D) governmental = service sector = higher education sector = the same = private non-profit sector = total financed by the private non-profit sector.

[2] This sector has been observed from 1991.

GERD= Gross Domestic Expenditure on R&D.

Table 3.2-3:   Biotechnology-related Contracts in OMFB Bidding System by
                 Sectors (1991-1996, January)

Sector	Number of contracts	Funding support (in million HUF)
Cultivation of plants	61	525.7
Animal husbandry	37	259.6
Food processing	2	21.7
Pharmaceutical	19	366.4
Health care	37	286.9
Environmental protection	16	11.5
Total	172	1417.8
All areas	1425	12769.0
Biotechnology-related as a percentage of all areas %	12.1	11.5

Source: OMFB data bank

Biotechnology, which was considered a high technology priority area, received
about 12 % of funding. This is not a small fraction for a priority area but it has to be
taken into account that private investment is on a very low level. In countries with
successful biotechnology business on the other hand the proportion of business is
substantial. The other, major problem is the size of funding: 1991-1996 funding
support per capita was 101.3 US$ and 50 % of contract financing are refundable.

The role played by government in the emergence and diffusion of this new technology is broader than the governmental programmes. The economic environment as a
whole can help or hinder its development. Besides financing, another important
issue for the biotechnology community is the legal framework.

3.2.2   Legal Framework

The general legal framework was summarised in the introductory part of this section. Now, there will be given details of only two legal framework conditions that
are particularly important for biotechnology:

- Regulations related to risks of biotechnology

- Intellectual property rights

Regulations Related to Risks of Biotechnology

In order to take full advantage of modern biotechnology, every country needs to establish regulations geared towards safety.[14] Setting up and implementing regulations in a timely manner will encourage innovation by providing planning security and will also help prevent negative impacts. The public will be able to utilise the benefits of the new technologies without incurring unacceptable risks to health, safety or the environment (UN 1992: p. 170).

In 1976 the National Institutes of Health (NIH) in the USA published safety guidelines for biotechnology, which later influenced the development of the respective OECD guidelines which were issued in 1986. The OECD guidelines have been used by many countries as a basis for developing national policy and procedures. Several directives were published by the European Economic Community (EEC).[15]
Hungary has not yet established procedures and criteria for evaluating the release of new products; however, her membership in the European Parliament requires to prepare and enact such regulations. The guidelines established by the United States National Institutes of Health, the OECD and the EEC guidelines and the recommendation of the European Parliament[16] may be useful. In the early 1990s the OMFB started to work on an Act concerning control, application, import and use of genetically modified microorganisms. This preparatory work was unfinished in 1994 when the status of OMFB changed.[17] Since then, three ministries - the Ministry of Welfare, the Ministry of Agriculture, and the Ministry of Environmental Protection - are preparing the act. Submission to Parliament is planned for autumn 1996. Parliament might enact it before the end of the year.

Intellectual Property Rights

From the aspect of patent law, the concept of biotechnology covers inventions involving the use and/or the production of a microorganism. According to the interpretation of the European Patent Office, biotechnology is control or modification of

---

[14] The first manual on biosafety was published in 1976 by USA National Institute of Health.

[15] Council Directive 90/219/EEC covers any contained use of genetically modified microorganisms, both for research and commercial purposes and Council Directive 90/220/EEC on experimental and marketing related aspects of genetically-modified organisms covers any R&D release of these organisms into the environment and contains a specific environmental risk assessment for the placing of any product containing or consisting of such organisms onto the market.

[16] The European Parliament's recommendation (PE-S-GG /90/5 concerning controlled application of use of genetically modified microorganisms) was translated into Hungarian but this translation has never been published and it was not on the agenda of any committees of Parliament.

[17] It is no longer a government agency.

the living organism by human intervention and for human use. (European Patent Law on biotechnology and related issues is discussed by Schmoch at al. 1992)

Among the intellectual property rights, five have relevance for biotechnology: pharmaceutical inventions; the protection of plant protective; biotechnological and genetic inventions; plant varieties and animal breeds; and deposit of microorganisms for patent procedure (See the details in Annex II.)

Generally, the Hungarian patent system was harmonised to international norms (EPO, TRIPs Agreement) by new laws which came into force on January 1st 1996.

3.2.3   Public Acceptance and Information on Biotechnology

Within the former centrally planned economy, technology assessment and public participation played a very limited part in policy and decision-making. The implementation of technology assessment in the present market-oriented economy has encountered a number of obstacles, such as lack of understanding of its importance in decision-making and a lack of suitable methodologies.

Public participation and public acceptance of a new technology is one issue of technology assessment. There is no general view on public acceptance of biotechnology in Hungary. Though it was decided to set up a Hungarian Technology Assessment Office, such an institution has not been established yet. Government agencies appear to be reluctant to finance research on technology assessment issues. Public debate on technology in general and on biotechnology in particular is not very well developed. In contrast to market economies where the media report, for example, intensively on the failures of biopharmaceuticals during clinical developments and where there is much discussion on health care, Hungarian media have not raised these issues yet.

The lack of information is the key reason why there is almost no public debate on the risks of biotechnology, uncertainties of its technological development or possible ill effects. Society does not exert any pressure on the government to establish biosafety institutions for the testing of transgenic organisms. There has been neither public controversy over environmental and human health concerns nor on socio-economic, ethical and regulatory issues. As mentioned earlier, there is no regulation on the introduction of genetically modified plants into the environment. But in laboratories there is some testing. For example, Biological Research Centre, Szeged tested five plants to see how genetically modified plants spread in environment. This was a co-operation project with other institutes.

In the next section the empirical analysis will illustrate that the low level of public information on biotechnology does not mean an easy environment for business, as might be expected. It has to be taken into account that present public acceptance of biotechnology is based on lack of knowledge and not on an informed assessment of this technology. Therefore, it is not clear yet how improved information on biotechnology will affect public opinion. It might as well cause favourable as unfavourable conditions.

3.3 Empirical Analysis of the Present Situation

The following sections analyse the findings of the empirical research. After commenting on methodological issues such as the aims of the audit and the sample of firms and institutes investigated, a broad outline of biotechnology activities and results will be given. Then the focus will be on different issues fostering or hampering innovation in Hungary. The innovation system will be analysed as well as strategies and resources for innovation on the micro-level. Then the business network which is an important factor of the marketing of innovations will be described. The previous discussion of different aspects determining the innovation potential lead to the concluding discussion of the competitiveness of Hungary in the field of biotechnology. The analysis of strengths, weaknesses, opportunities and threats will provide the starting point for the final evaluation and the formulation of policy strategies.

3.3.1 Audit as a Method

Overview

Technology audit is one of the tools to provide detailed and recent information for the elaboration of technological and industrial strategies. The first OECD review on Hungarian Science, Technology and Innovation Policies (1993) recommended to Hungarian authorities to undertake a thorough "technological audit". Its aim should be to identify technology areas with particular advantages based on strengths of the Hungarian innovation system, in which R&D capacities, industrial structures, educational basis etc. are important factors.

The audit should map the technological specialist knowledge and technology management knowledge and identify fields for improvement. In 1994, under the auspices of the OECD, four sectors were audited on a pilot basis by experts from OECD member countries (Germany, Austria, Finland and France). The selected sectors were: agricultural machinery, medical equipment (including laser equipment), plastic processing, and packaging.[18] This project showed that identification

---

[18] The Audit was carried out by four specialised institutions from OECD member countries:

of strengths and weaknesses through technology audit can be a first step in setting up priorities to support "the exploitation of technology knowledge".

Hungarian research institutions and other relevant institutions were involved and interrogated in the audit, but potential Hungarian evaluators were not invited to participate in the action. One important knowledge transfer possibility - learning by doing - was neglected. An international team offers a good chance of making international comparisons. By the end of the project this was replaced by bi-national comparisons, which is also important but is not the same evaluation level.

The pilot audit verified that it is a suitable instrument to support the process of transition in Hungary. It also clearly illustrated the requirement of methodological adaptation.

The present follow-up Technology Audit is a co-operation between two institutes: IKU Innovation Research Centre and FhG ISI Fraunhofer Institute for Systems and Innovation Research. Both institutes are independent of the government and without any business interests. The research team of the present Biotechnology Audit consists of national and foreign evaluators.

The bi-national research team - German and Hungarian - proved to be an appropriate combination. Hungarian experts are well acquainted with the situation in Hungary (Inzelt 1995a, b) and since Germans are not involved in everyday discussions on Hungary they can take a bird's-eye view of the country and they have more up-to-date information on world tendencies in the field of biotechnology. Another important characteristic of the mixed team is that members have different scientific backgrounds: biotechnology and economics. According to the experiences of the audit, scientific competence in technology and in economics (interdisciplinarity of experts) are key conditions for proper audit results. Collaboration characterised all phases of the process: defining the aims of the analysis, selection of application areas for biotechnology, developing interview guides, selection of sample, fieldwork, and analysis.

Experts from policy-making, industry and science were involved in the audit as interviewees and consultants. Contrary to the OECD Audit, appropriate Hungarian government institutions were not involved in this follow-up audit as evaluators. This made the investigation and evaluation of their role in the field of biotechnol-

---

Agricultural Machinery: The *Institute for Advanced Studies* (IHS), Vienna, Austria

Plastic Processing: The *Technical Research Centre of Finland* (VTT), Helsinki, Finland

Packaging: *Fraunhofer Management GmbH*, Munich, Germany

Medical Equipment: *Bertin & Cie.*, Paris, France and *Fraunhofer Management GmbH*, Munich, Germany

ogy possible. On the other hand, it was more difficult to carry out the audit and to obtain available data.

Our audit focuses on two categories of technology: technology already in use in the economy, and technology that is still at the level of laboratory prototype or undergoing development. This means that the Biotechnology Audit evaluates the capabilities and potential of sectors in which the technology is employed in order to help develop policies to target its development.

International comparison would be essential and it was one of the weakest points of the previous audits. In the present audit a survey of international activities in biotechnology is included which serves as a "yardstick" for comparison and evaluation of the Hungarian situation (chapters 2 and 4).

Methodological Problems

A great deal of information which is easily available in western countries is still lacking in transition economies. Consequently, the empirical study had to start at a very early information stage. The first step was to prepare a list of actors in biotechnology.[19] An up-to-date *register* is a general problem in transitions economies because of transformation and restructuring. It is even more complicated (not only in transition economies) to find a proper register in the field of biotechnology. The technology does not relate directly to specific sectors in the sense of statistical classification; rather, it appears in different sectors.

At the beginning of the investigation lists which contained organisations involved in biotechnology were collected.[20] These lists were complemented with press releases. For the preparation of the register, interviews were additional sources of information. Some of the persons interviewed mentioned organisations which were not covered by the lists. Annex II summarises the list of those interviewed experts, firms and institutes which agreed to publishing their names.

Analysis of Statistical Data

Without going into details of the information systems in central and east European countries, we have to mention the lack of technology audit-related information.

---

[19] Anikó Baricz, Györgyi Forray and Árpád Király were involved in preparation of this register.

[20] The following lists were available: owners of plant varieties, general R&D register at the National Technical Information Centre and Library (OMIKK), the lists of organisations awarded grants by the Hungarian Research Fund (OTKA), the lists of organisations receiving grants/supports from the Central Technological Development Fund (KMÜFA), the list of pharmaceutical firms registered by the Central Statistical Office, the lists of biotechnology-related engineering and scientific professional organisations, conferences. Dénes Dudits and László Heszky published a list in their book in 1990 on Plant Biotechnology.

There are very few useful data reflecting strengths and weaknesses in industry and technology compared with the data available in market economies. Even where data are available, their content differs from international standards. CEECs only have started to harmonise their measurement system with the OECD and EU. Similar to the OECD pilot Technology Audit, analyses at a macroeconomic level were not a primary goal of .this audit because many general macroeconomic analyses (e.g. OECD 1995a, b) were already available. For this follow-up audit, it was useful that the "Review of Recent Developments in Science and Technology in Hungary" (Inzelt 1995b) was available. This background study was a part of the OECD action and was prepared in parallel with the OECD Audit.

The OECD Audit concentrated on more mature industrial sectors, not on a very modern technology field. These traditional sectors are perceived as important for the economy and are more or less well covered by statistics. The follow-up audit covered a high tech area that is supposed to have potential in Hungary. As it was mentioned earlier in this case the statistical coverage is much weaker.

Journal-based Analysis

There are some biotechnology-related journals in Hungary. Apart from these, some industry-oriented professional journals are dying. The contents of professional journals were different from those in Western European countries. There were very few articles on applied research, experimental development or innovation. They usually discuss technical questions and publish new scientific results in the Hungarian biotechnology field. In these circumstances it would not have been reasonable to prepare bibliographic analysis; the scattered articles and theses supported the analysis and will be mentioned in the references.

Case Studies

Fresh case studies were not available. Some case studies concentrated on biotechnology in the 1980s (Frigyesi 1990a, b, 1993) and others on biotechnology-related industries, e.g. pharmaceutical sector (Antalóczy 1990, György 1993, 1994), agrofood business (Mohácsi 1996) in the 1990s. Beside a few articles, these mimeos and unpublished theses represented the sectoral literature studies. To a certain extent they were able to compensate for the deficits still existing in official statistics and interdisciplinary professional journals. During the investigation period we also prepared several case studies, two of which are included in part of this report as illustration.

Postal Questionnaire to Relevant Actors

There were not sent any postal questionnaires to those organisations that were not visited during the interview period. The reason was not only the time and cost constraints, but as mentioned above the incomplete register of biotechnology-oriented organisations in Hungary at the beginning of the audit actions. We had to start our work by compiling a list of institutes and firms that have some activities in the field of biotechnology. We were able to eliminate non-biotechnology actors from this preliminary list during the interview period.[21]

Expert Interviews

The number of experts who have an overview of Hungarian biotechnology is very limited. But interviews with experts were important information sources during different phases of the project: preparation phase (to test our idea, to ask advice on sample selection), main field work phase (structured interview with guide) and follow-up interview (checking some information collected and filling in some information gaps).

Interviews with Firms and Research Institutions

Similar to the OECD methodology concept, the core elements of the survey were interviews with firms and institutes and their own documentation. The investigation was conducted mainly in English. Most of the interviewees were fluent in English[22]. Most interviews were carried out by a mixed bi-national team from IKU and FhG ISI.

---

[21] Follow-up action would be useful to finalise this preliminary register.

[22] There were two reasons we prepared a few interviews in Hungarian: if there was no English-speaking person at the organisation interviewed, or only Hungarian researchers visited the firms/institutes. The interview guide was translated in Hungarian.

Table 3.3-1:   Outline of Interview Guides

Firms	Institutes	Experts

Overview
  Basic characteristics
  Business network
  Biotechnology activities
	Overview
	  Basic characteristics
	  Business network
	  Biotechnology activities
		Overview

Strengths and Weaknesses, Opportunities and Barriers
  Overview
  R&D
  Innovative activities
  Technology methods and processes
  Products
  SWOT Analysis
	Strengths and Weaknesses, Opportunities and Barriers
	  Overview
	  R&D
	
	  Technology methods and processes
	  Products
	  SWOT Analysis
		Strengths and Weaknesses, Opportunities and Barriers
		  Overview
		
		  SWOT Analysis

Competitive Position and Strategic Orientation	Competitive Position and Strategic Orientation	Competitive Position and Strategic Orientation

Recommendations
  Firms' actions
  Government actions
  Intermediary actions
	Recommendations
	  Firms' actions
	  Government actions
	  Intermediary actions
		Recommendations
		  Firms' actions
		  Government actions
		  Intermediary actions

Firm and institute management had open attitudes to the interviews. Participation was active in those firms that were willing to take part in the audit, similar to experiences during the pilot audit. During first contacts, a group of firms dropped out of the sample because they were in a pre-privatisation phase, re-deployment or had dropped biotechnology.[23] For the purpose of the audit this latter group of firms is considered to be less relevant, because they are not active in biotechnology and their future role in this field cannot be extensive.

---

[23] We made great efforts to learn the reasons why firms gave up their biotechnology activities. The phone interviews conducted during the sample selection process gave some explanations. Summarising the phone discussions it became clear why many firms that used to be involved in biotech (supported by Large-scale Biotechnology Development Governmental Programme) moved out of this field. Many of them had just finished experimental development by the end of the programme without any marketing or commercialisation efforts. Apart from the value of R&D results, there was no demand. In some other cases they had demand and they were in the commercialisation phase, but the new (foreign) owner evaluated biotech activity as less promising business and moved out of this field. An a few firms the R&D efforts did not have any real value. There are some examples of biotechnology disappearing from firms' portfolio because of spin-off companies.

The interview guides developed for this audit contained open and close-ended qualitative questions and close-ended quantitative questions. Table 3.3-1 shows the contents of interview guides.

The results of interviews were recorded in detailed interview guides. On this basis very comprehensive and detailed information is available.

One of the key points of a successful survey is to find the best *target person*, who could respond meaningfully to such a questionnaire. We targeted CEOs at firms, directors at institutes and heads of departments at universities. Their permission is needed to give information on organisations. In institutes and universities, the respondents usually were the directors/heads of departments; at a number of firms top management representatives were involved in the interviews. If a firm had an R&D director, he/she was the targeted person. Average duration of interviews was 4-6 hours. The interviewees could answer most of the questions readily. Certain answers required substantial research on the part of respondents. We sent them these questions in advance. This method improved the accuracy and consistency of answers.

The main lesson of methodological discussion is that the sample allowed to give an overall picture of the Hungarian biotechnology sector, even if written questionnaires were not used and some firms refused to give interviews. The study was able to cover the most important players, and contacted non-interviewed firms in some way too.[24] The following section presents the basic features of the sample. It then highlights the main findings of the empirical investigation.

3.3.2   Basic Features of the Sample

During the empirical study, 20 firms, 18 institutes and 8 experts were interviewed. Interviews with 17 institutes and 18 firms could be statistically analysed[25] (see the list in Annex III.). Several other organisations replied to only a few questions. This information and phone calls also contributed to understanding the changes during the transition period and to obtaining a better insight into the present situation.

---

[24] One group is missing from our sample: micro-firms around biotechnology. Our pre-field work concentrated on them. They are mainly users of biotechnology and many of them belong to this sector only in a very broad sense. We could identify many micro-businesses of plant varieties. According to our experiences they could be investigated with a much shorter, simplified guide.

[25] The reason for the differences between the number of organisations interviewed and the number in the data bank: some interviewed organisations gave important and relevant information but they were not ready to respond on some key questions.

We started to investigate biotechnology-oriented organisations in six sectors, namely: agriculture, chemicals, pharmaceuticals, food and beverages, diagnostics and others. There are no chemical firms devoted to biotechnology in our sample and diagnostics also has very limited importance in Hungary. In this field there are some user organisations and very few producers. Food and beverages are underrepresented in our sample because the owners (multinationals) usually did not agree to conduct interviews at their firms. Phone conversations indicated that some of them have introduced foreign-developed biotechnology products (e.g. milk industry), and some have closed down former Hungarian biotechnology activities (e.g. breweries).

The detailed analysis of the audit could concentrate on two sectors where biotechnology is applied: pharmaceuticals and agriculture. It is not by chance that there are only a few pharmaceutical firms in our sample and these are relatively small firms. Our sample is strong in agriculture. This sector was the key targeted sector of the Large-scale Governmental programme. This sector was able to survive regardless of turbulent changes in ownership, form of organisation, management etc. Large Hungarian pharmaceutical firms on the other hand have very limited commercialisation in biotechnology, R&D activities were diminished. Most of these firms are quitting biotechnology.

Because the Hungarian balance-sheet system is not establishment-based, very few firms and institutes could provide data on economic performance of their biotechnology business. They provided figures on their total activities and usually avoided giving rough figures on biotechnology. The following Table 3.3-2 gives some impression of size by turnover, export and R&D expenditure of firms.

Table 3.3-2:   Size of Firms by their Turnover and Export Earnings

Size categories (Mio. HUF)	Turnover		Export		Export to advanced countries		R&D expenditure	
	total	bio	total	bio	total	bio	total	bio
below 50	2	5	1	2	-	2	10	8
50-100	-	2	4	2	4	1	1	-
100-300	5	2	1	-	-	-	1	-
300-500	2	1	1	-	1	-	-	-
500-1,000	-	-	2	-	1	-	1	-
1,000-3,000	7	1	-	-	2	-	-	-
above 3,000	1	-	1	-	-	-	-	-
Number of respondents	17	11	10	4	8	3	13	8

The data show that firms involved in biotechnology had a turnover of between 13 and 5,200 million Hungarian Forints. The largest R&D expenditure related to biotechnology was 10 million HUF. In global terms, the firms are not large even if we calculate their total turnover. The average exchange rate was 125.69 HUF/US$ in 1995. This means that the largest investigated firm produced 79,560 US$/year. The income of institutes and university departments was between 0.6 and 600 million HUF. Incomes of university departments were below 10 million HUF. Only a part of this income came from biotechnology.

Table 3.3-3:   Employment Situation: Size of Firms and Institutes

Size categories (number of employees)	Firms						Institutes			
	Staff in total		R&D staff		Scientific staff	Technical staff	Staff in total		Scientific staff	Technical staff
	T	B	T	B	T	T	T	B	B	B
Below 10	2	1	8	7	11	6	2	6	8	10
11 - 25	2	5	3	2	2	2	4	3	4	1
26 - 50	-	1	3	-	-	-	2	2	2	2
51 - 100	1	3	-	-	-	-	5	-	-	-
101 - 300	6	1	-	-	-	-	2	-	1	1
above 300	5	-	-	-	-	-	2	-	-	-
Number of respondents	16	11	14	9	13	8	17	11	15	14

Note: T - Total; B - Biotechnology

The number of employees by size categories is shown in table 3.3-3 The total number represents the staff at organisations where biotechnology was a part of the research portfolio or the economic activity. Biotechnology data are estimates in those cases. There are no registered data on changes in employment. But every interviewed person mentioned a loss of scientists and massive withdrawal of technical staff.

Three investigated institutes were originally established before 1945 and four after 1981. The same figures for firms are: 4 and 7. The turbulent history of Hungary forced institutes and enterprises to transform themselves several times. The latest major transformation started around the beginning of the transition period and two thirds of investigated firms and a quarter of institutes were transferred into their present form after 1991. As regards the ownership, two thirds of the investigated firms are private. There are no private institutes in the sample. In the absence of a non-profit law, two institutes operate as quasi non-profit organisations.

3.3.3   Biotechnology Activities

In terms of research, development and technology biotechnology activities cover a rather broad spectrum in Hungary. Many different modern technologies are employed. *Analytical techniques* are used most frequently by firms and institutes. Firms use *biochemicals* and *computer/automation* techniques which support biotechnology processes. The profile of techniques used at institutes is different as they are deeply involved in *downstream processing* and *cell and tissue culture*. Among the biotechnology methods and processes *genetic engineering* ranks 5th at institutes, and 7th at firms in terms of frequency of users. Genetic engineering is not yet of high importance in the pharmaceutical industry. This represents a strong contrast to practice in advanced market economies. *Proteins* are used by institutes to much less extent, firms investigated do not employ proteins (see figure 3.3-1).

While differences in methods and processes employed in firms and institutes can be expected, firms make use of the knowledge available in institutes only to dramatically low extent. For example, institutes are employing DNA probes, sequencing and synthesis, protein sequencing and synthesis, and hybridoma technologies but these are not used by business organisations.

Another question concerns the organisations' own contributions to the methods employed (see table 3.3-4).

Figure 3.3-1:   Frequency of Used Biotechnology Methods and Processes

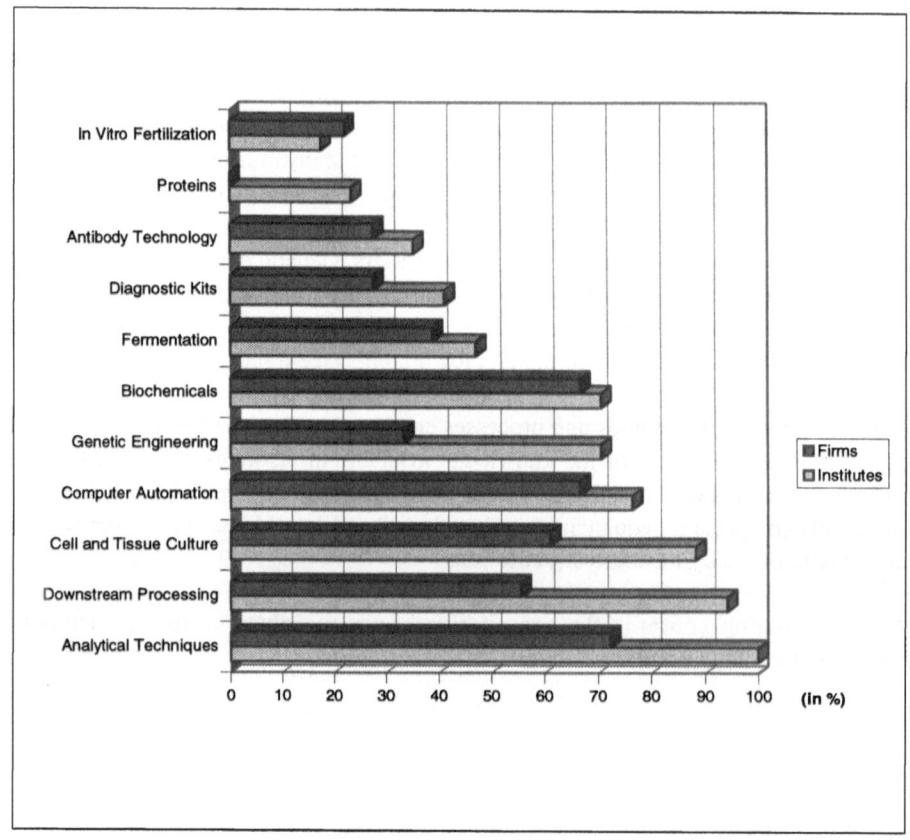

Table 3.3-4:   Own Contribution to Biotechnological Methods*

Methods	Share of own contribution		
	low	substantial	dominant
Institutes			
Tissue/cell culture	2	2	8
Biochemicals / reagents / peptides / enzymes	9	1	2
Computer hardware/software	7	3	0
Firms			
Microscopy	8	2	1
Computer hardware/software	8	1	1

* The table contains cases in which the number of respondents was 10 or more.

Figure 3.3-2:   Distribution of Products Targeted by Firms and Institutes
                 Respectively

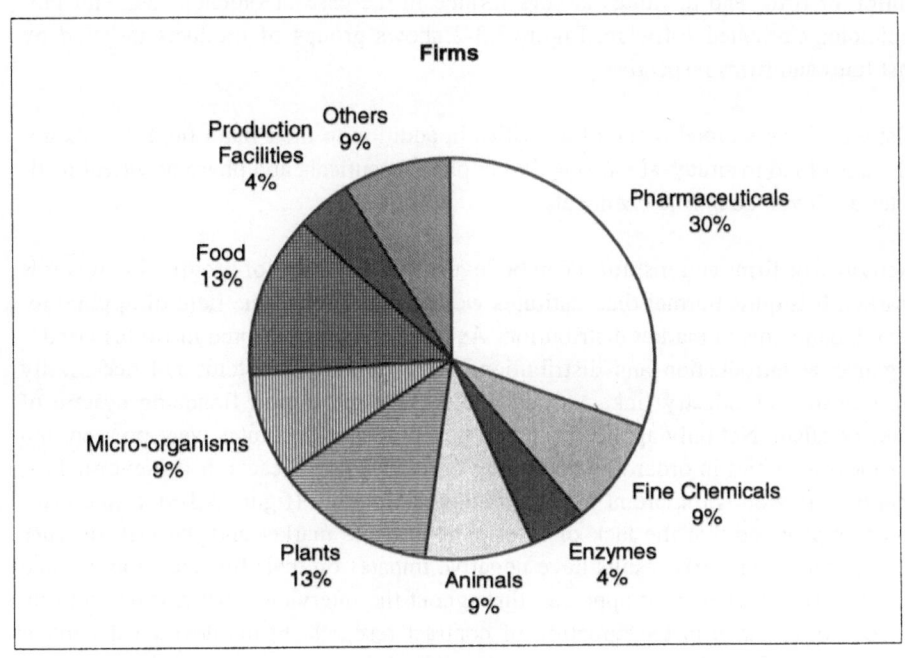

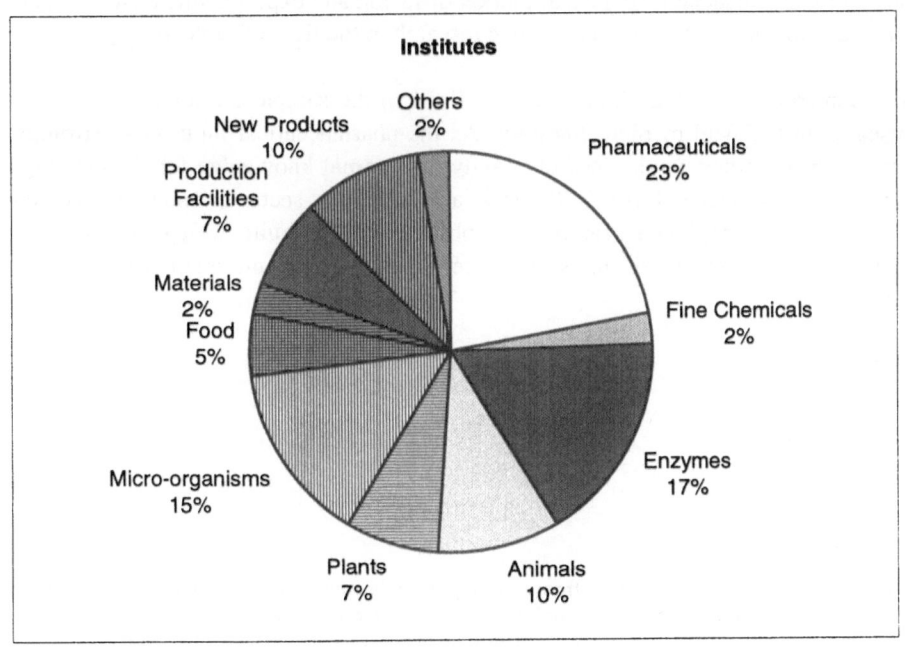

A dominant share of own contribution is found in tissue and cell culture. The Hungarian plant breeding traditions continue in the new technological culture. Contribution of firms and institutes is very limited in the case of biochemicals and biotechnology oriented software. Figure 3.3-2 shows groups of products targeted by institutes and firms respectively.

Institutes have a broad range of activities in addition to their main field. For example, agro-food institutes are also active in pharmaceuticals and pharmaceutical institutes are involved in fine chemicals.

Activities of firms and institutes can be investigated in terms of progression towards market. It is quite normal that institutes were most active in the field of applied research and firms in market distribution. As can be expected, some institutes are doing market introduction and distribution. This phenomenon stems not necessarily from academia-industry linkages but results from the hybrid financing system of late socialism. Not only applied, but also basic research institutes were pressed into business activities in order to earn money to finance basic research.[26] Nevertheless, market diffusion is important for knowledge distribution (figure 3.3-3). Some institutes emphasised that the lack of information on the market and the lack of sales organisations for R&D results have negative impacts on their financial sources and on the research agenda. It appeared throughout the interviews that researchers preferred not to move in the direction of contract research. Firms devoted the major share of their financial resources to market diffusion and experimental development. They conduct a limited fraction of basic research in the field of biotech.

It is remarkable that the pharmaceutical firms in the sample do not conduct basic research in the field of biotechnology. As the pharmaceutical industry is strongly science-linked, they would depend heavily on external knowledge for the development of new products. Firms in the food and beverages sector are neither involved in basic nor in applied research. This observation fits quite well to the general (international) tendency of this sector where mainly a few multinationals do intensive research.

---

[26] This "financial system" diverges from our subject. In the new legal environment institutes have terminated these profit-oriented activities. Apart from the fact that the new laws and rules are forcing them to restructure the confused system, demand for these small-scale high-value products has also disappeared as a result of liberalisation. The problem is that institutes' "self-support" was not replaced by public support or by business contracts.

Figure 3.3-3:   Activities Progressed to Market

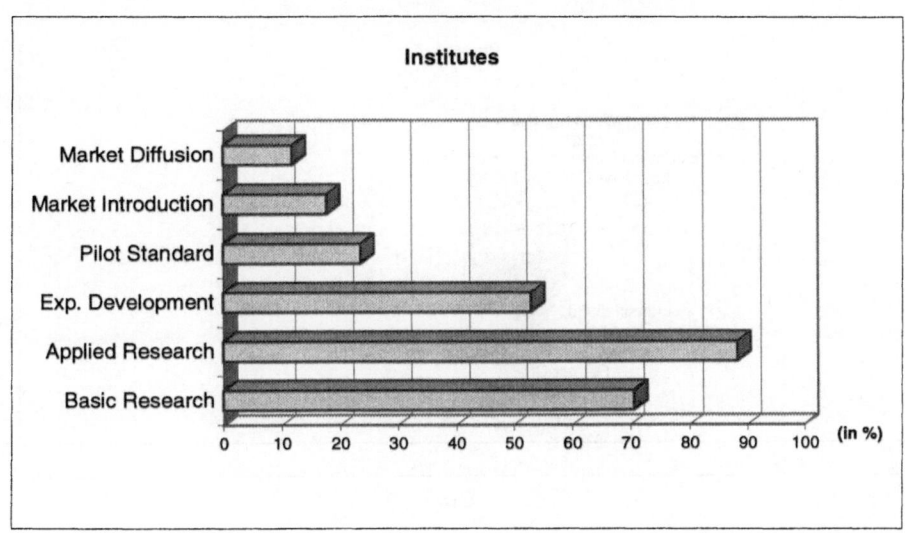

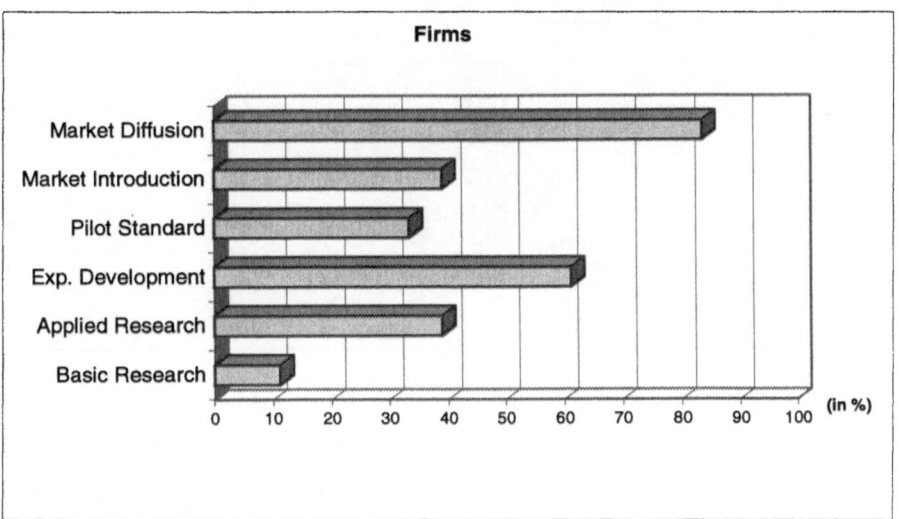

3.3.4   Market for Hungarian Biotechnology

Although the majority of Hungarian biotechnology products are still in the experimental and development stage, the analysis of present and future target markets gives an impression of the actual and potential commercialisation (figure 3.3-4).

Figure 3.3-4:   Target Markets of Biotechnology Products

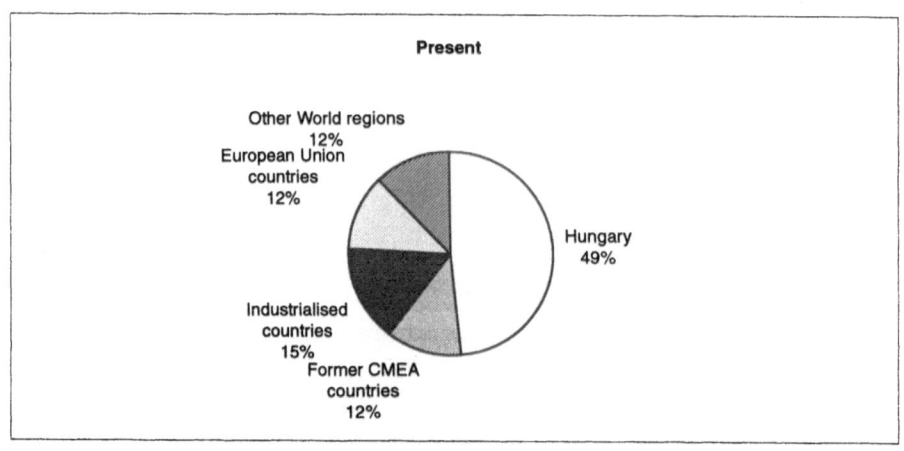

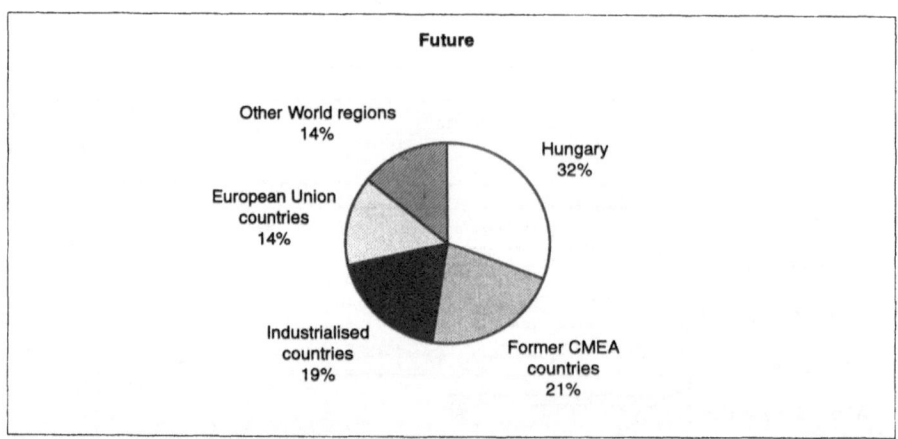

Recently, the domestic market has been the main target market for almost all respondent firms. Many of them lost their former foreign market in the last five years
because of the collapse of CMEA and shrinking possibilities for sales to developing
countries.[27] They hope to be able to gain export markets in the future and significantly increase their penetration into former CMEA countries and other regions,
too. In terms of market share the distribution of firms displays a two-humped shape.
In the domestic market their share is less than 5 % for 60 % of products and above
50 % for 40 % of products.

The latter are in a monopoly position on the domestic market. This monopoly position is, however, not strong enough because domestic demand is very weak in gen-

---

[27] Our sample includes only those that did not move out of biotechnology. During the preparation
phase we had many phone calls with firms that have terminated their biotech activities.

eral and is not associated with large profits. With the exception of one product, all firms' share is under 5 % on the world market. They have a better position in the non-European market than in the EU. Most of their products have a world market share less of than 1 %. Given that internationally there are many big players (especially in the pharmaceutical sector) with a rather low share on the world market (below 5 % in the pharmaceutical sector), the situation of Hungarian firms is not very different. Several firms in the sample do not have any information on their share of foreign markets.

Marketing is weak not only at the institutes but at a large number of firms, too. Institutes are not informed about the research demand of their potential clients and they have only limited knowledge on their small-scale production market. In many cases companies which will be privatised shortly do not have any market strategy. Privatised companies usually have a short-term market strategy; some of them are testing new targeted markets while others are not aware of the importance of marketing. Therefore, new products designed for existing demand do often not lead to economic success. For example the lack of customers' awareness of biotech products and their advantages leads to only limited demand (e.g. lactose-free milk, virus-free plants, test-tube calves).

There are some preconditions to broadening the market share in old markets and to competing in new markets. Both firms and institutes considered that they have to overcome *financial bottlenecks for marketing*. They mentioned a *non-existent or weak sales organisations* as the second great burdening factor. The third hampering factor was a *non-existent or unsatisfactory market know-how* in the case of institutes, and too expensive products application procedures in the case of firms. The institutes do not know the demand for their R&D portfolio and are still expecting too much information from the government.

3.3.5   Innovations

The present section analyses the functioning of the national system of innovation. Despite scientific achievements the innovation capabilities of the economy seem to be weak. Innovation is a precondition for economic growth and maintaining employment and can contribute to achieve competitiveness in the long run. Innovation depends on the application of scientific results in industry leading to marketable new processes and products. Therefore, the creation and diffusion of new technologies involve dynamic interactions between firms and their environments. Important elements of the innovation system are networks between suppliers and customers, support services, financial institutions but also the publicly supported infrastructure for science and technology including universities and R&D programmes.

In the following, the Hungarian system of innovation will be analysed. The firms' and institutes' evaluation of the former and present organisation of R&D in Hungary will be taken into account (chapter 3.3.5.1). Then, the innovation performance of enterprises and institutes will be investigated on the micro-level.

3.3.5.1 Innovation System

The majority of respondents judged the former model of R&D organisation negatively despite its possible advantages. The interviewed persons concentrated on the *financial issue* as the key factor of the organisation of the research system. Only one respondent considered the supporting system of the former model, which financed institutes without evaluation and competition, superior to the present one. Most interviewees criticised this aspect because it did not encourage the research community to produce achievements. The main problem of the present model is its transitional character that co-exists with a sharp decrease in both public and business funds. The sudden decline in financial resources is the reason for many interrupted or unfinished projects. The bidding system which underlies the two large public funds (OMFB and OTKA) was evaluated positively although the shortage of money signifies an obstacle to its effective functioning. Some enterprises saw the need to concentrate the funds available. Biotechnology firms emphasised the role of financing for R&D and innovation, non-financial measures are considered as less important. Summing up the perspective for business-financed projects does not appear to be very promising. Enterprises tend to wait for new scientific results for commercialisation but expect government to cover R&D expenditure at institutes.

Beside this general bottleneck of the system a special biotechnology-related aspect was mentioned: there are no proper laws governing the Hungarian system. The absence of a *legal framework* for biotechnology has two negative impacts: (1) there is a real danger that some companies will carry out research projects in the country which are not allowed abroad; (2) in the uncertain legal environment institutes are reluctant to invest in new projects to avoid losses associated with a forthcoming law which would be too rigorous.

Industry demands the creation of *transfer organisations* and regular forums between academia and industry. No one mentioned his own or professional associations' tasks in this process; recommendation were made towards government. From the point of view of *industry, R&D portfolio organisations* are problematic. Firms highly appreciated basic research at a few institutes but co-operation would be more useful if institutes showed more willingness to be involved in experimental development, testing and consulting.

Diffusion of biotechnology depends on many factors, from highly qualified, properly funded research to commercial potential. For policy-makers it is important to identify problem areas in this system in order to be able to target respective policy measures. We asked firms and institutes to evaluate 15 problem areas (table 3.3-5).

Table 3.3-5:   Rank of Impeding Factors to R&D in Biotechnology

Factor	Rank order at	
	institutes	firms
Bottlenecks in R&D financing	1	2
Uncertainty about market, lack of marketing knowledge	2	3
R&D expenditure too high in view of anticipated economic success *	3	1
Inadequate technical equipment	4	4
Public knowledge of technology	5	8
Inadequate public promotion	6	5
Too few suitable co-operation partners	7	10
Acceptance of biotechnology by public	8	15
Procurement of information on technical development trends	9-10	6-7
Patent situation	9-10	13
Uncertainty about development path adopted because of the importance of competing technologies	11	6-7
Legal framework conditions for biotechnology	12-13	11
Technology field difficult to grasp because of its complexity *	12-13	9
Approval practice/law implementation	14	12
Lack of technological know-how in the firms	15	14

Note: *less than 12 respondents

The first four factors are the same at institutes and firms: *bottlenecks in R&D financing, uncertainty about market, lack of marketing knowledge, R&D expenditure too high in view of anticipated economic success, inadequate technical equipment.* Ranking is clearly different by firms and by institutes in the case of *public acceptance of biotechnology.* Firms do not consider this aspect as a serious problem. Differences in the ranking of other factors are also related to the significance of these factors for firms or institutes respectively, e.g. *uncertainty about development path adopted because of the importance of competing technologies* was in 6th or 7th place at firms and ranked only 11th at institutes.

The environment for innovations can support or hamper the diffusion of knowledge, co-operation, market penetration etc. In the following, elements of the macro-economic system that are preconditions for improved dissemination of knowledge, innovation and competitiveness in the field of biotechnology will be analysed.

(1) Financial resources are very scarce, even more for innovative products. To name only a few problems, public finance is limited and uncertain, the tax environment for investments in innovations is unfavourable, the willingness to incur risk in the early stage of innovation is weak. For companies, many potential partners (banks, pension funds, venture capital firms, business angels etc.) hardly exist as financiers of innovations. Besides this, to evaluate the opportunities and risks associated with biotechnology is a great challenge for financing systems and instruments even in advanced market economies.

(2) Interregnum in policy formulation, lack of foresighted actions.

(3) There persists an inadequacy of legal and regulatory environment which has a transitional character in many fields. Some special regulations for biotechnology are still missing.

(4) Physical infrastructure. Both firms and institutes emphasised that some elements of existing Hungarian technical infrastructure are bottlenecks to innovation, e.g. the lack of telephone lines or access to e-mail are burdening factors with regional differences. When compared to the recent past, the situation has changed significantly in a positive direction. There are far more telephone lines, and digital exchange centres have improved the quality of lines considerably. The research community can freely use e-mail and computers are commonly used tools. All communication infrastructures are available but their relative cost is higher than in advanced market economies.[28]

(5) Intermediary institutions, associations, and chambers are lacking or hardly working (see section 3.3.6).

(6) Research is weakly oriented towards marketable innovation.

(7) Co-operation between the domestic scientific community and the business sector is weak. Hungarian firms are involved in international co-operation only to a very limited extent.

3.3.5.2   Innovation in Firms

Research and development are essential components of innovation. Hungary devotes 1 % of her GDP to R&D. This is around the level of less developed small OECD countries. Industrial research carried out and financed by businesses is on a small scale. In-house expenditure on R&D including biotechnology is very limited (firms usually could not provide figures on their biotechnology R&D expenditure, but very few of them spent anything on it). Spending on in-house research is important, even if firms do not like to produce new scientific results. According to inter-

---

[28] The number of mobile telephones is relatively high because both business organisations and individuals were willing to invest in them to overcome bad public telephone conditions.

national experiences, if a firm does not have any R&D capability it is hardly able to adopt (acquire) new technologies.

Inputs into Innovation

The *expenditure* on innovation and the breakdown of costs (R&D, acquisition of patents and licences, product design etc.) can characterise the inputs depending on the firms' strategy. The main goal of firms' R&D efforts is *product and process development* (13 respondents). This target was followed by gaining competence with *monitoring and following competitive trends* (4). Only one firm mentioned *reaching a strong patent position* as its own goal.

Improving the capabilities of firms is also a key factor of innovations. Present and expected *technical capabilities* within two years are shown in table 3.3-6.

Table 3.3-6:   Existing and Expected Capabilities

Denomination	Leading edge	Modern	Used modern	Completely outdated
Firms				
Technology standard				
present	7	3	3	3
future	9	6	3	-
Engineering				
present	6	5	4	2
future	7	7	3	-
Plant equipment				
present	6	3	6	2
future	7	8	3	-
Institutes				
Technology				
present	-	10	7	-
future				

University-industry linkages are very weak. All institutes emphasised that their R&D programmes are determined by science, not by industry or policy. Only one institute mentioned that its R&D programmes are determined predominantly by the market.

This paragraph investigates the importance of different *R&D output of scientific organisations*. Publications are the most preferred results, bibliometric methods became an important instrument in the evaluation of researchers and institutes in the late days of socialism. This is associated with positive and negative consequences.

If the future of researchers strongly depends on one factor, they will overemphasise its importance. For example, to publish in a periodical with 4 to 6 impact factors is much more important for many of them than applying for patents. Patent application requires postponing publication.[29] If we make a distinction between institutes and university departments, publication efforts are much less important in the latter group in biotechnology according to their self-evaluation. Most departments have only limited experiences in the field of biotechnology, mainly for educational reasons. Some departments accumulated good research capabilities, but in the last few years leading researchers left departments because of uncertainties in the general financing of universities. At the same time, some university laboratories became well-equipped thanks to different international programmes (e.g. FEFA). The key problem is that only investment costs were covered and after the expiring of the grants there are no resources for recurring costs[30].

The quality of Hungarian R&D personnel is evaluated highly. This is an important source for innovation. However this asset cannot be valorised without other inputs.

Concerning the size of inputs in biotechnology innovation the critical mass for economic success is missing at every stage.[31] Furthermore, the time introduction of a new product into the market is becoming a crucial factor in competition. If the capabilities for these are missing, industry will have less chance for success. In the last decade biotechnology research became much more cost intensive than it was in the 1970s. If basic research organisations cannot cover the research costs in time, much less first-rate scientific result will be achieved in Hungary.

The Main Objectives of Innovation

Firms ranked their objectives for innovation in 1996 as follows (table 3.3-7):

---

[29] This negative attitude is strengthened by bad experiences with using patents, collecting royalties etc.

[30] In winter 1995 and 1996 several university buildings were closed for weeks because they did not have resources to cover heating costs. Phone calls and faxes are limited.

[31] This is not only a problem of firms but basic research laboratories, too. As one of the young researchers stressed, they need two years for laboratory tests instead of three months because they do not have enough money to repeat a day test even if they have equipment. Lots of researchers are willing to work day and night to speed up laboratory test phases.

Table 3.3-7:   Ranking of Objectives for Innovation in 1996

Objective	Number of firms marking objective
Increase (or phase out) market share	11
Improve product flexibility	10
Extend product range within bio-technology	9
Open up new markets abroad	8
Replace products	6
Lower production costs	6
New domestic target groups	5
Other aims listed below	5

These priorities reflect typical survival strategy and aspirations of firms are more wishful than operational. In the long-term perspective improving quality was mentioned as the most decisive element of their strategy. The second aim is to improve the price/cost ratio and the third to penetrate into market niches. Innovation activity (which means in this context the development and introduction of new products or processes) comes only after these strategical issues.

Key Sources of Information for Innovation

Among the inputs, the sources of information are very important. Conferences and journals were of key importance for more than half of the responding firms. Second came educational and research institutes and third in-house R&D. While these are relevant sources, perceived market needs which finally will decide on the economic success of innovation are not considered by enterprises. Networks are not explored as an information source. They have not yet reached the building up period. Respondents showed interest in obtaining information about possibilities for partnership and co-operation. Acquisition of embodied and disembodied technologies, licences, clients, or suppliers has low importance. As possible sources of information acquisition of innovative firms and investment in innovative technologies are not employed.

The range of used sources of information provides empirical evidence that there is almost no feedback between different stages in the innovation process.

Missing Players for Innovations

With the exception of a few sectors (e.g. horticulture), small biotechnology companies are missing. There are only very few spin-off companies. As discussed above (chapter 3.2.1), the underdeveloped innovation financing system is one of the key impeding factors for spin-off formation. A few spin-off companies are typical phenomena in the transition economy (see one example in Box 1).

3.3.5.3 Patent Activities

Patent indicators are useful for analysing economic and technological relationships, and R&D output. Table 3.3-8 illustrates total patent applications in Hungary since the beginning of the transition period.

Table 3.3-8:   Number of Patent Applications in Hungary

Applicants	1989	1990	1991	1992	1993	1994	1995
Hungarian	2,657	2,506	2,199	1,629	1,186	1,153	1059
institutional		964	738	541	347	341	312
individuals		1542	1461	1088	839	812	747
foreigner	4,278	6,044	8,379	8,291	10,828	16,535	19449
direct		2038	1827	1690	1407	2640	2870
PCT		4006	6552	6601	9421	13895	16579
temporary protection						(60)	(692)
Total	6,935	8,550	10,578	9,920	12,014	17748	21200

Source: Hungarian Patent Office, OECD (WIPO)

Patent applications in these years show two different tendencies. First, applications by Hungarian inventors are declining, probably because economic restructuring has lead to closing down of companies, dismantling R&D departments, changes in the patent system, and declining R&D expenditure. Second, patent applications by foreign investors have risen sharply, a good indication of the internationalisation of the Hungarian market (Inzelt 1996a).

The first patent applications of biotechnological subject were submitted to the National Office of Inventions (new name: National Patent Office) in 1978. In the first decade of modern biotechnology about 1000 patent applications were submitted to the Hungarian Patent Office.[32]

Most of the applicants were non-residents, mainly multinational companies from advanced economies. This application activity clearly shows that foreign companies are interested in the Hungarian market. It also indicates that the national intellectual property rights environment was acceptable for them. The main resident applicants were pharmaceutical companies, some research organisations and individuals.

---

[32] Biotech related patent data were compiled by Dr. Mária Petz-Stifter.

The Patent Office was reorganised in 1989 to respond to changes in the structure of invention fields. A new department (Biotechnology and Agriculture) was set up and they had to recruit new evaluators with competence in biotechnology. Exact information on patent activities in the field of biotechnology has been available since 1989.

Prior to the reorganisation of the Patent Office the first application arrived from the Regents of the University of California, San Francisco in 1978 (table 3.3-9). The first Hungarian organisations among the applicants were the Institute for Drug Research five years later in 1983. This was the thirties application on recombinant biotechnology. Following Hungarian applicants were Chemical Works of Richter Gedeon Ltd., one of the largest Hungarian pharmaceutical firms and the most famous basic research centre in Hungary, namely the Biological Research Centre of Hungarian Academy of Sciences, Szeged. It is worth mentioning the Hungarian Biotechnika Share Co. (1984), a new company established for commercialisation of biotechnology research results. This company did not appear again among the filers until 1990 and went into bankruptcy. Such an organisation presumably was premature in the past Hungarian socio-economic environment.

There are also some individual patent holders among the filers. At the time of registration in the 1980s these individuals worked at large Hungarian pharmaceutical firms and their patentable results were not covered by their job contract.[33]

---

[33] Many such patents are characterised as weak patents. Firms usually did not try to introduce them, or license them out. The reason for their origin and non-exploitation was different from market economies. It was rooted in the socialist wage system where companies could hardly differentiate among their employees by their capabilities and usefulness. The IPR system made it possible to compensate somehow the best engineers, designers etc.

Table 3.3-9:   First Thirty Filers of Recombinant Biotechnology in Hungary

Applicant	Rank	Country of Origin	Date of Application	Date of Registration
The Regents of the University of California	1	US	26/05/1978	28/04/1987
	3		30/05/1980	28/10/1988
	10		25/08/1981	30/05/1988
Biogen N.V., Willemstadt, Holland Antilles	2	NL	20/12/1979	28/10/1988
	5		08/01/1981	28/01/1988
	8		01/04/1981	28/10/1987
Schering AG	4	DE	01/08/1980	28/10/1987
CPC International Inc. Englewood Cliffes	6	US	13/02/1981	30/05/1988
	12		15/01/1982	28/10/1987
Genentech, Inc., South San Francisco	7	US	23/03/1981	30/05/1988
	11		09/11/1981	28/10/1987
	16		30/08/1982	28/10/1988
	20		19/10/1982	15/10/1990
	23		07/03/1983	24/07/1991
	25		14/04/1983	29/11/1990
	27		04/05/1983	31/11/1990
F. Hoffmann-La Roche Co. - Genentech Inc.	9	CH - US	29/06/1981	28/11/1988
Hoechst AG	13	DE	01/06/1982	28/08/1987
Boehringer Mannheim GmbH	14	DE	03/06/1982	28/09/1987
Eli Lilly and Co., Indianapolis	15	US	17/06/1982	28/04/1988
	19		14/10/1982	28/10/1988
	21		25/11/1982	28/02/1989
	26		15/04/1983	28/02/1989
	28		23/05/1983	28/03/1989
Orion-Yhtyma Oy	17	FI	29/09/1982	28/10/1988
Ciba Geigy AG	18	CH	01/10/1982	28/02/1989
Asahi Kasei Kogyo Kabushiki Kaisha	22	JP	23/12/1982	28/11/1988
Schering Corp., Knilworth	24	US	14/03/1983	30/01/1989
Dr. Karl Thomae GmbH	29	DE	27/05/1983	28/11/1988
Institute for Drug Research	30	HU	11/07/1983	28/02/1989

Source: Patent Office Department of Biotechnology

The number of pending biotechnological applications amounts to about 1000, of which 20 to 30 relate to transgenic plants. Until 1995 the Hungarian Patent Office did not receive any patent application concerning transgenic animals. In the field of recombinant biotechnology, the most active filers at the Hungarian Patent Office were the following (table 3.3-10):

Table 3.3-10:   Ranking of Domestic Patent Holders (at the Beginning of 1996)

Rank	Number of Patents	Name of Patent Holders	Type of Holders
1	9	Biotechnology Research Institute of Szeged, HAS	R&D Institute
2	7	Chemical Works of Richter Gedeon Ltd.	Firm
3	6	Reanal Share Co	Firm
4	5	Institute for Drug Research, Budapest	R&D Institute
5	3	Agricultural Biotechnology Research Centre, Gödöllő	R&D Institute
6	3	Phylaxia Serum Producer Co.	Firm
7	2	Chinoin Share Co	Firm

Taken together, the patent figures presented in this section show that there are remarkable patent activities in the field of biotechnology in Hungary.[34] Patent statistics can measure to a certain extent the results of R&D efforts, but cannot give information on the "strength" of patents. The scientific value of a patent is only one important aspect of its expected benefits. The use of a patent depends on many economic factors including characteristics of global competition. A patent is a prerequisite as well as an expression of expected economic success. It depends on the national economic environment, the economic potential and strategy of inventors/filers whether they will use the patent, license it out or just leave it as a sleeping patent. In the following section the economic output of R&D activities, use of patents etc. will be investigated.

3.3.6   Business Network

By its nature innovation is an interactive process between the actors of the innovation system. In the context of the Technology Audit we try to investigate business relations between their major partners: clients/customers (companies or industries), suppliers, different actors of the public sector, financiers, alliances, R&D partners and associations.

---

[34] R&D organisations and academic researchers are not very active in patenting. The main problem is that they usually do not have enough financial resources to develop their research results into a patentable product or process. If they can do it they also need funds to cover the registration fee etc.

We questioned not only the firms but institutes too on this issue. The importance of players was ranked by number of respondents and frequency of each relationship. For example, if a firm had permanent relations with a partner it was evaluated as more important than if it had only occasional relations. Table 3.3-11 shows the result of this weighted evaluation.

Table 3.3-11:   Importance of Partners

Relations	Firms	Institutes
Government	1	1
Authorities	2	8
Banks	3	7
Suppliers	4	5
Association	5	4
Customers	6	6
R&D partners	7	3
Agencies	8	2
Consultants	9	9
Alliances	10	10
Other financial institutes	11	11
Others	12	12

Not only the ranking of partners differs between firms and institutes but also the frequency of their relations for the government. The government (as key financier) is a more critical partner for institutes than firms. R&D partners do not have the highest importance among partners. This is not surprising because they were not among the key sources of innovation (see section 3.3.5.2). Only four of the investigated firms have permanent contacts and three mentioned occasional contacts. Others did not mention any relations in this field. This empirical evidence is strengthened by institutes' experiences. Institutes are waiting for firms as partners to finance their research projects and experimental development.[35] If firms would like to remain players on the biotechnology stage it would be advisable to consider R&D partners more seriously.

In the following a brief survey by partners is given:

Government was mentioned as an important financier usually by hospitals because they receive regular budget support. But this relation is weakening as Health Insur-

---

[35] Agricultural institutes are usually ready to meet firms' demands. They are client oriented research organisations. But they also have to face the problem of a lack of demand. Some basic research organisations e. g. university departments declared they are ready to continue applied research or experimental development on a contract basis to earn money to develop their laboratories and to receive normal wages.

ance is becoming an increasingly important financier. Pharmaceutical firms emphasised that they used to have strong relations with the government (via the Ministries of Health and Foreign Trade). Now this relationship has been replaced by Health Insurance, which determines their domestic prices. Agricultural firms mentioned the Ministry of Agriculture as an important source for information on biotechnology and economic issues. Some of the firms interviewed are still owned by the state. For them state (relevant governmental organisations) is important as owner.

Departments of universities have critical relations with the Ministry of Education which is responsible for the allocation of their budget and manages the Educational Fund. Institutes receive their financial resources from the Ministry of Education. Beside this, agricultural institutes can get support from the Ministry of Agriculture. Among the critical relations governmental funds were mentioned: OMFB as manager of KMUFA fund, by firms and institutes, OTKA by institutes and universities, FEFA by universities. The character of the relationship illustrates well how business and science are moving away from the bureaucracy. Research agenda and business projects are not determined by the government while there is some influence on organisations via the budget allocation.

All investigated organisations have regular relations with *authorities*. These are compulsory relations because e.g. of registration requirements of products and partly research. [36]

Banks' main role is to handle accounts and transfer money. Hungarian banks were creditors in a few cases. They are crucial because it is extremely difficult for firms to obtain capital from other sources. Joint ventures or 100 % foreign-owned firms usually work with foreign banks. One joint venture mentioned that the relationship with banks could be crucial as owners are not willing to invest in new technology. It is worth noting that only one firm mentioned it has relations with a *financial institute* other than a bank. This financial institute was not a domestic one. Relations were occasional. This is further evidence of the inadequacy of the financial market.

Agencies (chambers, councils of products, technology transfer agencies): the respondents emphasised the changing character of *chambers*. They appear to be busy with self-reorganisation and can hardly support members. Some of them are consid-

---

[36] (Agricultural Quality Control Institute, Agricultural Qualifying Institute /Mezõgazdasági Minõsítõ Intézet/, Veterinary Hygiene Stations of the counties /Megyei Állategészségügyi állomások/, Food control Station /Élelmiszer Minõseg Ellenõrzõ Állomás/, State Public Health and Medical Officer Service, Public Hygiene and Epidemiology Service /ÁMTSZ/KÖJÁL/, National Food and Nutrition Institute /Országos Élelmiszeripari- és Táplálkozástudományi Intézet/, National Institute of Pharmacy /Országos Gyógyszerészeti Intézet/, Scientific Research Ethical Committee, National Institution for Public Health /OKI/). Some pharmaceutical firms have relations not only with Hungarian authorities but with foreign bodies, too. For example the US Food and Drug Administration and similar inspection agencies in EU countries.

ered over-bureaucratised. The expectations towards agencies are high, for example, firms rely on chambers helping them in strategy formulation in the future. Councils of Products have an important role for some agricultural firms but these activities are not related to biotechnology. The firms considered that these organisations have hardly any influence on the success of their biotechnology activities.

Institutes have relations with many *associations*.[37] Firms are also members of several associations but their relations are less frequent.[38] One institute is not a member of any Hungarian association but it is affiliated to an international professional association.[39] Associations are important for maintaining and developing professional contacts and exchanging information. According to institutes, most associations are responding to members' needs but some are over-bureaucratic and especially exchange of information could be improved.[40]

Most of the institutes have *R&D partners*. Partnership usually means exchange of information, joint research and in some cases co-operation in education. The lists of partners shows two important features: (1) many institutes co-operate with domestic partners which in general is not very common in Hungary. Domestic partners are other departments of the same university/institute, or other universities/institutes. (2) Remarkably, institutes and universities have many foreign co-operation partners.[41] Not only the number of foreign partners is impressive but many of them belong to leading institutes and organisations in biotechnology worldwide (e.g. DFG, FhG, University of Göttingen, University of Cologne, Max Planck Aquacul-

[37] E.g. National Associations of Fish Farms, Biochemical A. (Associations), Biophysics A., Pharmacological A., Breeders Animal Husbandry A.

[38] E.g. A. of Vegetable Growers, A. of Hungarian Farmers, A. of Horticulture, A. of Hungarian Meat Market, Medicine Herb A., Hungarian Federation of Safferers from Coeliac Disease/, A. of Diabetics. Hungarian Assisted Reproduction Society, A. of Holstein-Friesian Breeders.

[39] European Embryo Plantation A., World Embrioplantation A.

[40] Leaders of professional associations usually used to work at research organisations but they have lost solidarity with the scientific community. (This means that old routine can survive the decay of a system in which associations represented government and not their members. This phenomenon is interesting because the elected leaders have changed.)

[41] Hungarian economic reform (1968) has had some positive impacts on international relations of the business and scientific communities. Biotechnology did not belong to any large CMEA programmes, (last S&T co-operation programme just declared as a target without any enforcement) In this field co-operations started with advanced market economies and developing countries. Before the transitional period there were some R&D co-operations, visiting fellowships, and few partnerships on the basis of Hungarian basic research results with MNCs (Eli Lilly, Pharmaceutical Research Institute Ltd.) Business co-operation also were established on the basis of individual contracts permitted by government. (Japanese TAKEDA-Chinoin /pharmaceutical firm/, Nádudvar /state-owned farm/ and Japanese firm /now it is a joint venture: Agroferm Hungarian-Japanese Fermentation Share Co./).

ture Institute (Germany), Texas State University, University of Columbia (United States), Museum of Natural Science (France), Hampshire College (Great Britain)).

Institutes and universities have some co-operation with firms. Two types of contract research are most common: joint research projects with firms and R&D contracted out by firms. One institute which used to work on contract projects for Hungarian pharmaceutical firms was able to replace this lost relation with a foreign partner. Hospitals have co-operation with universities and multinationals (e.g. Akzo, Hoffmann-La Roche). One investigated private hospital has strong R&D relation with its foreign owners.

Few firms have R&D co-operation or commission research projects. One enterprise owns a share of 30 % of a so called research company. Besides many formal relationships they have extensive informal relations with Hungarian universities and institutes. One is working together with a British institute. Some firms also have scientific and experimental development co-operation with MNCs e.g. Bayer, Eli Lilly, Ciba-Geigy, Du Pont Pharma, Merck etc.[42]. As a new phenomenon worth mentioning, Hungarian private laboratories are among the partners at some agricultural firms.

The significance of external co-operation is summarised in table 3.3-12.

The purchase of foreign R&D was not considered to be an issue. Both, firms' and institutes' technological co-operation with domestic partners was much more important than with advanced market economies. Former CMEA countries have limited roles in technological co-operations.

Firms and institutes have regular *suppliers and clients*. The biotechnology sector is very sensitive to suppliers because it has the respect of health and safety regulations by end-producers who strongly depend on raw materials and intermediary products. Many of the suppliers are foreign companies (e.g. Sigma, Merck, Serva, Pharmacia, Beka, Hewlett Packard, Organon (Dutch), Ferring (German), Serono (Italian), Medicult (Danish), Memezo (French)). One reason is that several export markets can accept only those products whose suppliers were strictly selected. Another important fact is that these large firms can offer much lower prices than former Hungarian producers because of scale effects. Liberalisation of the market allows companies to choose among potential suppliers. Some new Hungarian suppliers are on the list with small scale special products, e.g. Bay Zoltán Institute for Biotechnology, József Attila University, Szeged, Nyírő Gyula Hospital, Haynal Imre Health Science University. Among these organisations are no small firms but non-profit-

---

[42] It is not easy to obtain correct information on the content of these co-operations. The firms mentioned together tableting, storing and marketing drugs on the Hungarian market, and managing registration, analysing the medicines and R&D.

oriented institutes, universities and hospitals. Some agricultural firms have subcontractor relations to strengthen the connection with their frequent partner.

Table 3.3-12:   Importance of External Co-operation in R&D

Denomination	Institutes		Firms	
	important	very important	important	very important
Technological Co-operation	8	6	8	7
- with domestic partners	9	4	4	6
- with partners from advanced market econo-mies	5	1	1	2
- with partners from former CMEA coun-tries	-	1	-	1
- with others	-	-	-	3
Purchase of For-eign R&D	-	-	-	-

Universities and institutes in general have a much less constant circle of suppliers. They have many occasional relations. Their domestic suppliers (farms, chemicals) are rather stable and their relations are very frequent.

Consultant or, consulting firms play hardly any role in the business network of firms. It was not mentioned as an external source, but subsidiary companies regarded their parent companies as important and frequently used consulting partners.

3.3.7   Competitiveness

The Biotechnology Audit tried to identify appropriate actions to improve the competitive position of Hungarian biotechnology in the business and in scientific arenas. The SWOT (Strengths and Weaknesses and Opportunities and Threats) analysis provides detailed starting points for strategic action. This section analyses the competitive starting position of firms and institutes from which their biotechnology business can develop. These initial conditions provide the basis for the formulation of a strategic vision and has to be taken into account when developing an action plan for implementation. If this strategy exceeds the firms' resources or ignores their weaknesses, additional external actions (e.g. governmental action) are required to realise the strategy. The first part summarises the results of the SWOT analysis. Then the patterns of the organisations' strategic orientation and crucial factors for success will be identified.

3.3.7.1   Strengths and Weaknesses and Opportunities and Threats

The SWOT analysis[43] was conducted as a joint evaluation of interviewees and interviewers in order to make a proper distinction between strengths and opportunities weaknesses and threats.

Internal Factors

All interviewees regarded the *quality of human resources* (including researchers, engineers, technicians) as a strength of their organisation (table 3.3-13)[44]. Most of the organisations consider this not only a present strength but also an opportunity for the future. But some of them fear that the quantity of qualified researchers and supporting staff may become a constraining factor because institutes are not able to offer adequate salaries to retain qualified staff.[45] The same reason explains why the recruitment of more research and technical staff is not possible. The lack of personnel is a weakness at almost all organisations and will be a strong barrier for future development. PhD programmes and special supporting programmes are evaluated as a good chances. It is a strong threat that highly qualified employees will leave hospitals, firms and institutes because of low wages. The low cost of human resources offers very limited, temporary possibilities. Knowledge and capability of human resources were highly appreciated. Some firms mentioned the importance of training and re-training to improve their future possibilities. The main reason is that none of the investigated firms was fully devoted to biotechnology, so they have very few people with adequate knowledge. To integrate biotechnology into their companies there is a requirement to re-train people (see Box 1).

Generally, low Hungarian *R&D costs* are considered as strengths of organisations, but *low salaries of R&D personnel* signify weaknesses of institutes and universities. There are two other general weaknesses at institutes and universities: *lack of marketing knowledge and capabilities* and *weak management*.

---

[43] It is a part of some postgraduate programmes. This question was very much welcomed by respondents who felt summing up critical strengths and weaknesses to be an extremely useful exercise.

[44] As evidence institutes mentioned publications in prestigious journals, numerous patents, regular co-operation with foreign laboratories.

[45] It has to be emphasised that low salaries do not mean advantages for organisations, contrary to the widespread myths. In the field of biotechnology there are very limited possibilities for professional second and third jobs.

Apart from low expenditure on R&D six firms mentioned R&D activities as their strength. Four of them are able to use their own staff to develop new products and processes. Two others have good R&D co-operation with other firms.

Box 1

Biological Research Centre of Szeged

The Biological Research Centre (BRC) is one of the youngest institutes of the Hungarian Academy of Sciences, it started to work in 1971. The foundation of the research centre coincided with the scientific revolution which resulted in the birth of modern biotechnology. The BRC has worked with the most modern biological techniques and methods since the beginning. The official opening ceremony was conducted by the Nobel prize winner Hungarian-born biochemist Albert Szent-Györgyi. The first director of the BRC was F. Brúnó Straub, who was the most successful disciple of Professor Szent-Györgyi.

The research centre consists of 5 Institutes of Biophysics, Biochemistry, Enzymology, Genetics and Plant Biology. These institutes have the same administration, library, workshops etc. The different institutes are concerned with basic research, their research areas belong to the fields of molecular and cellular biology, such as studying the role of one of the key protein in inhibiting the AIDS virus, studying the protective mechanism of the tumor cells, examination of the effect of ozone on plant photosynthesis, examination of the stress reaction of animal and plant cells. Many groups of researchers have got remarkable results. Ádám Kondorosi and his group was one of the first in the world to start working with symbiotic nitrogen fixation in the 1970s, and he became internationally famous and a respected expert and researcher of the mechanism of nitrogen fixation. Now he is the director of the Plant Biotechnological Institute of the French National Research Centre, the CNRS, and also continues working with his group at Szeged.

There are 445 employees of the BRC, 225 of these are scientific staff (1995). After the decrease in the last years, the number of the scientists may remain at this level. In the Centre a traditionally large number of students and Ph.D. students are working at the moment, too. The number of the Ph.D. students is growing permanently; it was 55 in 1995. The Biological Research Centre has participated in all domestic biotechnological education and all the education of the foreign young scientists since starting its work. The senior researchers usually give courses at the Attila József University of Szeged and the BRC has 15 foreign young scientists from different developing countries (1995), who learn the modern molecular biology in the framework of the International Training Course financed by UNESCO.

The main financial resources of the institutes come from the Hungarian Scientific Academy, different project applications (EU, PHARE, OTKA, OMFB), and R&D co-operation with industry. More than 50 % of their income originates from state support, about 20 % from international public sources (EU, PHARE), about 10 %

from contracts with firms (for example Ciba-Geigy, Hoechst) and a small share is from foreign private funds (Volkswagen Foundation) occasionally. About 12 % of the research projects of the BRC obtains grants from the state by OTKA and OMFB (1995).

The BRC has many troubles to contend with, which are based on the precarious financial resources. The good start in the 1970s was due to the high state support and the Developing Programme of United Nations. This helped to purchase the modern equipment, to invite first-rate foreign experts to launch the new research projects and to send their young scientists to leading laboratories in western Europe and the United States, to learn the new techniques and methods. As a result of this help and the fact that there was no similar institute in central eastern Europe, the BRC became a "professional centre" in this region.

The financing of the BRC has been changing unfavourably from the beginning of 1990s: the costs of the research centre increased very much while the real value of its income decreased. Most of the equipment became "used modern" and this standard will fall within the next 2 years. Access to literature is becoming a new additional problem, because the central library cancelled many journals due to financial restrictions.

The main R&D partners of the BRC are among the institutes and the universities. The BRC is working together with foreign research institutes in more than 50 % of the research projects. The Hungarian institutes and universities share in 10-20 % of the BRC projects. The most important foreign partners are from Germany, UK and USA. The BRC has the most research contacts with the Agricultural Research Institute of the Hungarian Academy of Sciences of Martonvásár, Agricultural Biotechnological Research Centre of Gödöllő, Medical Herb Research Institute, Food Industrial University of Szeged, Semmelweis Medical University, and the Attila József Science University of Szeged of the Hungarian institutes and universities.

The scientists of BRC publish 150-200 items in highly respected international journals per year. In 1988-1990 120 articles were published in journals with impact factors above 4. The R&D outputs of the institutes appear only occasionally in other forms, as for example patent registration, technology transfer out, requests for technology partners.

BRC has 9 registered patents (1996), 6 were registered abroad, too. Among these are some prominent patents, for example one was worked out by the Plant Biology Institute and claims to be a more efficient method for plant biotechnological gene transfer. We should take this patent into consideration because it was worked out by more actors: the Corn Research Institute of Szeged, the Hoechst Company, and the Ciba-Geigy Co.

The BRC is co-operating also with Hungarian companies, primarily with pharmaceutical firms, secondly agricultural firms. They have contract research for example with Chinoin and Richter Gedeon Chemical Works, State Horse Farm of Mezőhegyes Sh.C., "Hajós" Grape and Wine Farm of Szeged. From the co-operation with the Richter Pharmaceutical Company also a patent was born, but it did not become commercially successful, because the production was not realised until today.

The following factors are considered as strengths of the Biological Research Centre: low R&D expenditure compared to similar leading European institutes, researchers have high expertise, they have favourable opportunities to co-operate with domestic and foreign institutes and there are a lot of young inspired scientists, Ph.D. students with a strong sense of vocation.

Compiled by Anikó Baricz

Commercialisation is not a strength of the organisations investigated. Institutes have developed results that are premature for commercialisation. A common problem is that many research results are in the pre-commercialisation phase until they become obsolete[46]. Scale-up capabilities are weak at institutes and firms. Because of *rare R&D co-operation* knowledge development and acquisition is suboptimal. There were some exceptions in the sample: strong university-clinic linkage is a guarantee of introducing scientific results into practice; an experimental institute with its own production line can commercialise its own R&D results in a short period although it has to accept lacking economies of scale. In the case of firms, one mentioned as its own strength fast commercialisation capabilities, another uses world-class biotechnology processes. A few others emphasised, competitive products as their strengths. Prices and product portfolio also support competitiveness. One firm highlighted its vertical integrated production chain as a strength.

Equipment is a very controversial issue. Some institutes have up-to-date modern equipment and also in sufficient quantity. Some others have good equipment but less than they need. Several university departments have high quality equipment, but they cannot cover the running costs to use them. Most firms emphasised as their strengths the good quality of instruments which are suitable for further development. In a few cases, out-dated equipment is the greatest weakness of institutes and firms. Two firms are on the threshold of further growth: they have fully exploited their present biotechnology capacities. Without investment they are not able to increase their production and exploit potential markets. One possibility is, as some firms mentioned, to invest in equipment for growing and introducing new products. But it is also a threat in those cases where investment is too risky or there is strong dependency on the parent company in investment, licences etc.

---

[46] Some Hungarian patents were quite good at the time of their registration, but they lost their potential profitability without commercialisation.

Table 3.3-13:   Self-evaluation of In-house Factors

In-house factors	Impacts		
	weak	average	strong
R&D potential			
institutes ·	4	9	4
firms	8	6	3
Technology			
institutes	2	8	5
firms	2	5	10
Product			
institutes	7	5	1
firms	0	6	9
Equipment			
institutes	1	0	1
firms	4	7	7
Marketing/sales			
institutes	13	0	1
firms	8	5	3
Finance			
institutes	13	1	1
firms	6	9	2
Suppliers, logistics			
institutes	1	1	0
firms	2	9	6
Human resources (quality)			
institutes	0	7	10
firms	2	6	9
Salaries			
institutes	8	2	5
firms	2	10	5
Management			
institutes	6	7	1
firms	1	11	5

Half a dozen firms have a *high level of technology* which is a good basis to develop their possibilities: to develop more biotechnology products and processes, to broaden the biotechnology portfolio, to use more biotechnology methods, to improve technology and products and to avoid threats, to increase the gap between them and their competitors. A few firms are successful with a monocultural technology or product respectively. If they cannot introduce others they are threatened by declining demand, decreasing prices and profit.

Good marketing and *good partnerships* are a strength of few firms. A significant share of the domestic market (5 firms) and large product capabilities as the basis for flexibility are also considered as strengths (4 firms). Good relations with co-operation partners were mentioned by three firms as strengths. This is very realistic because as other surveys have shown, firms do not explore the opportunities of learning by co-operation. Marketing is a weakness at many firms. Many enterprises do not have a clear concept of marketing, market shares and the formulation of appropriate market strategies. Without the improvements in the field of marketing, the threats are very realistic: they will not be able to adapt to changing domestic market and will respond too slowly to world market challenges. It is very difficult to avoid strong dependency on one or a few clients. Some other firms do not only consider weak marketing as a threat but rely on possibilities to improve it and regard good marketing as a key factor of their future success. Their strategy to develop and improve marketing capabilities is to find marketing specialists who can identify the potential demand for the enterprise's products and support the employment of existing capacities. It may allow them to exploit their existing strengths (products, processes, human resources) much better, to increase their market share and penetrate new markets.

As table 3.3-13 shows, *management* is regarded as a strength at more firms than institutes. The common characteristic of well managed firms is that they were privatised and foreigners are among their owners. At those firms where management belongs to the weaknesses, they emphasised that they are not modern in their thinking because they were educated in the old system and their behaviour and routines originated from there.

Size of R&D staff is small in many cases and is a weakness of many organisations. Only three firms and four institutes evaluated their own *R&D potential* as a strong factor. Nine institutes and six firms characterised it as average, and four institutes and eight firms thought their potential is weak. Weak has two meanings: there are not enough staff to concentrate on time on R&D, or R&D is only a side job of product engineers and technicians. Low R&D potential might be a serious problem for firms' future. Success or failure of adaptation depends on in-house R&D capabilities. Among the possibilities, some recovering firms mentioned that they may set up R&D departments, establish more R&D co-operation with foreign partners, and with domestic universities and institutes. At the same time firms will be threatened to lose their own R&D capabilities if they cannot invest more. Lack of experiences in R&D co-operation is also a threat to establishing proper co-operation in a short time.

According to firms' self-evaluation, their *technology* is strong. Institutes are less satisfied with this factor.

External Factors

Among external factors (table 3.3-14) institutes and firms considered the *bottleneck of financing biotechnology* as most important hindrance for biotechnology development.[47] Governmental support is shrinking in real terms. Public funds offer only limited support to basic research projects and financial means available are uncertain. The industry sector has not recovered yet, and enterprises are reluctant to cover R&D costs. Only one firm declared as its strength its stable financial position. The financial market is not developed enough to finance experimental development and the market introduction of biotechnology activities. The financial situation, lack of capital and excessively high tax rates, high interest rates, devaluation of the Hungarian currency and high inflation represent a burden of investment and a threat for the future of Hungarian industry.

Given the general economic situation and arising uncertainties, unclear national economic policy and regulation, and short-term thinking of economic actors during the transition period, the future development of biotechnology is threatened. For the pharmaceutical sector, hospitals and the diagnostic sector, the reform of health care insurance regulations might cause decline in demand and dissolve existing good relations. This could lead to the expansion of MNCs which are also their main competitors on the domestic market.

There are advantages and disadvantages too that a special *biotechnology law* has not yet been enacted. There is no lobby against biotechnology in Hungary which can be considered a strength and opportunity. It would be possible to avoid the emergence of a negative public attitude toward biotechnology if public knowledge was improved by appropriate information strategies.

The legal environment: imperfect legal regulation, many changes in the system lead to problematic framework conditions for firms. Furthermore, rules for biotechnology which are slow and costly impede the authorising procedure in the pharmaceutical and also in the food industry. Unclear regulation of foreign trade was also mentioned as an unfavourable factor.

---

[47] Financing is also an internal factor. But respondents concentrated on it only as an external factor. This is not surprising because they need fresh capital from outside if they want to develop their biotech activities.

160

Table 3.3-14:   Evaluation of External Factors of Competitiveness

External factors	Impacts		
	weak	average	strong
Geographical position in Europe			
institutes	2	9	5
firms	2	8	5
Local infrastructure			
institutes	2	10	4
firms	1	13	1
Industrial environment			
institutes	9	3	3
firms	6	9	3
Domestic market			
institutes	6	4	1
firms	5	5	5
Public funds and subsidies			
institutes	12	4	0
firms	13	2	0
Public regulation			
institutes	2	10	3
firms	7	8	0
Public acceptance			
institutes	3	8	4
firms	5	5	6

Insufficient demand for biotechnology products and processes are burdensome factors for institutes. Relations between science and industry are very weak, and co-operation within the industry sector are not very frequent either. Some firms feel the continuing lack of potential co-operation partners to be a real danger for their future.

Among the opportunities, mainly institutes but also firms emphasised new opportunities *to join European programmes,* until now access is rather weak. Some institutes evaluated their limited contacts with foreign business organisations as a weakness.

An external strength for organisations is the availability of *professional information.*[48] This may be very natural for Western European institutes and firms but it

Availability means there is no political/bureaucratic obstacle to obtaining journals, books, attending conferences, mobility programmes etc. Newly set-up information centres provide information on projects, funds etc. But they still have financial obstacles: basic research organisations often

used to be a seriously burdening factor for organisations in socialist economies. Because of this reason many respondents emphasised the information factor as a strength.

Basically, the self-evaluation of organisations was realistic, even if some opportunities are more an expression of wishes than real opportunities. Evaluation is usually based on their knowledge of the Hungarian situation. For example, physical infrastructure was overvalued because respondents compared it to the recent past and Hungarian average, not to the western European level. Some evaluations on behalf of institutes reveal restricted perspective but some have a good perception of their actual status. Mostly, these latter are integrated into the international scientific community and have a good background for evaluation.

Confidence in Strategic Analysis

In Hungary, strategic management encounters considerable problems because of unstable market and overall economic conditions: strategy formulation is very difficult, and analyses are usually not well founded. Several firms have to cope with problems of day-to-day survival, therefore, strategical thinking is very remote from the need to react to the changing environment. Few firms are aware of the importance of strategy formulation.

The SWOT analysis helped to prepare the self-evaluation which is the basis for the decision-making process and the strategy formulation.[49] I was emphasised that exploiting the results of the SWOT analysis depended strongly on financial conditions.

Referring to the SWOT analysis and possible strategic choices, firms' objectives are, for example, reduce costs of biotechnology products by new technologies, extend biotechnology product range introducing significantly changed products, increase or maintain market share and improve and stabilise quality of products. This shows the willingness to overcome in-house factors which hamper innovation. Enterprises would aim at strengthening their R&D capabilities either by re-establishing or increasing R&D activities to develop new equipment and new types of products or by carrying out joint research with domestic scientific groups which are competent in animal biotechnology.

---

have to cancel subscriptions to periodicals because of the increasing prices and decreasing support.

[49] One firm mentioned that they knew this method from their study tour to the US, but never used it. After this joint action with interviewers they may do so.

3.3.7.2 Competitive Position and Strategic Orientation

International comparison is of considerable significance for the evaluation and development of a high tech sector. This audit is the first attempt to put Hungarian biotechnology on the world map. However, tools for this exercise are very limited. Patent statistics shown in chapter 3.3.5.3 are one of the sources which can be used for this purpose. In the following sections this information is complemented by an evaluation of the Hungarian position by the organisations participating in the audit. This positioning takes into account the technological and economic situation. During the next step the competitive comparison is elaborated at the micro-level, namely firms and institutions evaluate their situation with respect to their main competitors.

Figure 3.3-5:   Biotechnology Position of Hungary in Relation to Groups of Countries (1-Developing Countries; 2-Tiger Countries; 3-Other Industrial Countries; 4-Leading Countries)

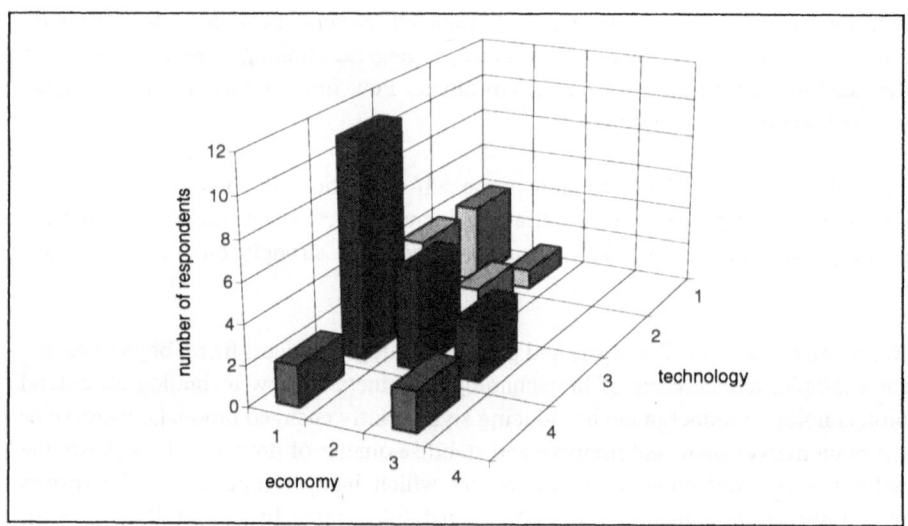

A few respondents (firms and institutes) evaluated the Hungarian economic situation (in biotechnology) as similar to that of non-leading industrialised countries (5) (figure 3.3-5). A higher share of respondents perceived Hungary close to tiger countries (8) and most of them believed the country's position is comparable to developing countries (21). None of the agricultural respondents ranked the country in terms of her economy into the group of non-leading industrialised countries. The picture looks different if we look at the place of the country by technology: four evaluated it as similar to leading industrialised countries, nineteen to other industrialised countries, six to tiger countries and only five to developing countries.

The main players on the world market are multinationals with a strong research base and strong marketing capabilities. In this context, biotechnology programmes at firms and institutes might not be realistic if compared to research expenditure that institutes and companies in advanced market economies can afford. The shortage of financial sources is a fact and the balance between wishes and possibilities cannot easily be found. The question is not only shortage of money but a critical mass of financial sources. Hungarian pharmaceutical companies appeared to be very realistic when emphasising that they cannot employ these success factors because of missing basic conditions. (In the opinion of one of them, biotechnology industry will not develop in Hungary because the technology is too expensive for smaller, less advanced countries).

Firms estimated the scenario for the future development of Hungarian biotechnology. All of them emphasised the importance of more expenditure on R&D and acquisition of licences. In these processes they expect more support and investment from the government, at least credits with guarantees, low-interest rates, a tax break period etc. Business organisations with the exception of pharmaceutical firms consider financing of R&D mainly a governmental task. Firms would only have to supplement these state expenditures if results are available. They attribute much more importance to frame conditions (e.g. government tasks) than to any other factors.

Out of this perspective government could play a more important role than simply being a good financier. For example, the coordination of government supported R&D activities (avoiding the dissipation of governmental support)[50] could be a task. Others include the investment in basic research and training in biotechnology. Firms emphasised as their own tasks developing new products and processes and patenting them.[51] The firms stressed that institutes and universities have to respond more to business demand (for example, one of them evaluated them as "bad partners" because they are not able to keep to time schedules). Permanent relations between university and industry could help to restructure their research portfolio. Apart from many professional organisations which were mentioned in section 3.3.6, some firms recommended setting up forums where industry and academia can meet and exchange ideas regularly.

---

[50] There are two different problems relating to the dissipation of governmental support: a small sum is divided among too many institutes and they work on the same topic without real competiton, evaluation etc. The other problem is that the size of year-by-year support is uncertain. There are too many semi-finished basic and applied research results. These are dormant values. If the dormant period is long they will produce no commercial results either for science or for business.

[51] Patenting has become a more important issue since the beginning of transition. Firms have had to face the importance of such assets. At companies privatised by MNCs the transfer of knowledge in and out of Hungarian subsidiary is a delicate issue.

In the field of biotechnology it is not conducive to monitor just local trends because world trends determine competitiveness. Tools and procedures for analysing the competitive position and strategies for development are not employed extensively. A part of them are the same as sources of innovations (see section 3.3.5.2) Other tools include: contracting out market research, evaluating sales data and the quality of products of the competitors, analysing clients' and customers' opinions, visiting fairs and exhibitions. Enterprises and institutes expect basic information from the Central Statistical Office.

The overall evaluation of the firms' position in relation to competitors on the world market (figure 3.3-6) indicates that most of the respondents consider themselves presently as technology followers and with respect to market as situated between a follower and a leader. The future position is expected to improve a little with respect to technology but not to market.

Before going into a detailed evaluation of their positions, it has to be mentioned that the knowledge of the firms about the international competitive situation is in many cases limited: only few firms could mention names of world's best and main competitors. This response rate again highlights the limited knowledge of the market and the weakness of this type of self-evaluation.

Figure 3.3-6:   Firms' Position in Relation to Competitors on the World Market

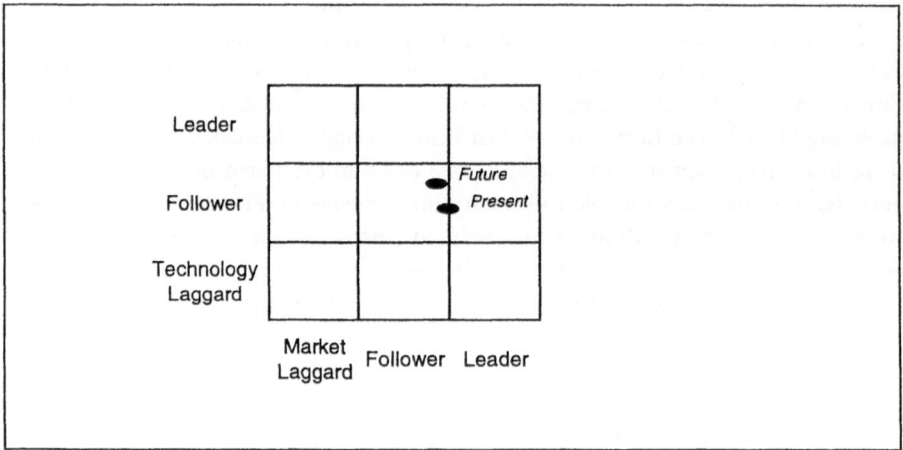

If we look at R&D resources and output figures in comparison to the main competitors they give some explanation of the present and future position. As table 3.3-15 shows government support for R&D expenditure is very low in the international comparison, mainly because of the very limited financial resources. Access to European Union programmes is higher in institutes than in firms. Partnership in EU programmes is important for transition economies not only because of the extra financial sources but also as a source of information and as a possibility to learn by interactions. Some factors were evaluated as "average" to main competitors. In the case of access to *literature, innovation infrastructure, conferences, scientific networks* and co-operations this estimation rather means significant improvement compared to the former position than having achieved already the average of their main competitors. The relatively best competitive position is seen with respect to R&D qualifications and know-how. This again underlines the high quality of human resources in Hungarian biotechnology.

Table 3.3-15:   R&D Resources in Comparison to Main Competitors

Resources	very low	low	average	high	very high
R&D expenditure	*14*	10	5	1	0
R&D manpower	7	6	*10*	7	2
R&D qualifications, know-how	1	5	8	*14*	2
R&D equipment	2	5	*16*	8	1
Scale and scope of own R&D	3	9	*13*	8	0
Financial resources	11	*16*	6	0	0
Access to literature	0	4	*12*	9	6
Access to innovation, infrastructure: tech. patents, associations, patent info, consultancy	0	8	*15*	4	4
Access to conferences	1	4	*15*	9	4
Access to info services	0	4	*18*	7	2
Access to scientific networks	3	5	*15*	6	1
R&D co-operation with domestic universities	5	3	*6*	5	0
R&D co-operation with domestic institutes	5	5	*8*	1	0
R&D co-operation with domestic companies	2	*6*	5	1	0
R&D co-operation with foreign universities	*9*	6	2	2	0
R&D co-operation with foreign institutes	8	6	*9*	*9*	0
R&D co-operation with foreign companies	7	11	*13*	2	0
Government support for R&D	*16*	13	4	1	0
Access for EU programmes	*13*	9	9	3	0
Other factors	1	1	*2*	1	0

Table 3.3-16:   R&D Output in Comparison to Main Competitors

Output	very low	low	average	high	very high
Publications per capita					
Institutes	0	3	6	7	2
Firms	6	4	6	1	0
Patent Position					
Institutes	7	9	0	1	1
Firms	10	1	2	2	0
Technology Transfer out of					
Institutes	1	6	7	3	1
Firms	8	3	4	2	0
Request as Technology Partner					
Institutes	1	3	10	5	2
Firms	3	2	5	6	2

Turning to the comparison of R&D-output, table 3.3-16 clearly shows that publications per capita indicate a relatively good performance of institutes and even of firms. But the patent activities are on a low level as well as technology transfer. On the other hand most firms and institutes evaluated themselves as attractive partners for technology co-operations.

4.   Comparison of International and Hungarian Development of Biotechnology

The aim of this chapter is to compare the Hungarian situation in biotechnology to world standards and recent developments. A set of criteria which is derived from the international experience is used as a yardstick to evaluate Hungary's performance. Therefore, within the first part of this chapter, the key factors for success in biotechnology will be identified. In the second part the Hungarian situation as described in chapter three will be analysed and evaluated against this framework of success factors. This is the basis for identifying institutional and individual elements of the Hungarian innovation system which could influence the described key factors or the Hungarian performance with respect to these factors. This analysis will be carried out in the concluding chapter 5.

4.1   Success Factors for Biotechnology - Lessons from the International Experience

Success factors for competitiveness can only be derived by taking into account the complex set of conditions which influence innovations in biotechnology (Reiss 1996). These conditions relate to

- technology-born features,
- industrial structures,
- conditions of the production system,
- broader framework conditions.

The most important **technology-born features of biotechnology** are a very strong science linkage, high innovation dynamics, multidisciplinarity and a modular production structure.

Science linkage means that development and application of biotechnology depends directly on results of related basic research. Therefore, a strong **research base** and highly skilled **manpower** can be identified as success factors. The science linkage of biotechnology can be measured, for example, by analysing to what extent new patents draw on knowledge which has been published not in previous patents but in scientific papers (Grupp and Schmoch 1992, Schmoch et al. 1993). Such analyses show that the science linkage of biotechnology is about three times higher than the science linkage of all technologies on average (Schmoch 1995).

The **high innovation dynamics** of biotechnology is indicated for example by the number of patent applications over time: patent applications in biotechnology at the European Patent Office have increased by a factor of five between 1980 and 1990 (Schmoch et al. 1992).

The **multidisciplinarity of biotechnology** is twofold: on the one side biotechnological innovations are a result of many different disciplines of science and engineering working together. On the other side biotechnology opens up new applications in several different sectors. A consequence of this network integration is that even if considering just the scientific and technological aspects of biotechnological innovation processes, it is obvious that usually not a single factor but many different factors and the quality of their interaction must be in an optimal condition in order to provide a competitive system. The transfer of this interdisciplinary know-how is an additional challenge for innovations in biotechnology.

The **modular production** structure of biotechnology describes the fact that biotechnological processes are performed within small units like, for example, bioreactors. More complex technical systems are built-up by the combination of different small units. Therefore, many interfaces between different modules and other technological areas are created, calling for system integration management.

In most countries the **industrial structure** of biotechnology is characterised by the prevalence of small firms with less than 50 employees. Among these small firms a rather high percentage belongs to the very small category with less than ten employees. On the other hand, especially in the health care sector, big multinational companies are important players in biotechnology. Within these firms biotechnology is usually organised in smaller units, which are comparable from their size with typical small and medium-sized firms. Most of the small biotechnology firms are rather young firms which have been founded during the last ten years.

Among the **conditions of the production system** the long time-to-market period is a peculiar feature of biotechnological innovations. At present the average time of bringing a new pharmaceutical to the market is about twelve years. Due to the application of new technology, this time span is being shortened to roughly eight years. This is still long enough to cause on the one hand high technical risk and on the other hand requires appropriate financing strategies. The technical risks associated with these long time-to-market periods are illustrated by the fact that out of 10,000 drug candidates finally only one to two will reach the market. The costs of developing a new drug can amount up to 500 million US$. From this, the important role of **market-orientation** and favourable **financing** conditions becomes clear.

Broader **framework conditions** for biotechnological innovations include mainly legal, social and ethical aspects. In all industrialised countries a certain framework of legal conditions has been developed which on one hand aims at assuring the safety of biotechnological activities. On the other hand, these frameworks should provide planning security for industrial activities in biotechnology. A third feature of the legal framework is the positive correlation between legal framework conditions and public confidence and acceptance. This in turn is a prerequisite for consumer willingness to accept new products of biotechnology. Property rights are another important component of the legal framework. International harmonisation is an important requirement for the protection of biotechnological inventions.

To summarise, the international experience with innovations in biotechnology (see Chapter 2) indicates that there are certain **success factors** which form the basis for meeting the challenge of the described key characteristics of biotechnology:

- research base,
- manpower,
- know-how transfer,
- interdisciplinary networking,
- market-orientation,
- financing,
- legal framework.

4.2 Performance of Biotechnology in Hungary in International Comparison

4.2.1 Research Base and Manpower

Research Base

Due to the strong science linkage, a broad and high quality **research base** is needed. As the American and the British examples indicate, basic research in molecular biology plays a key role in this context. These activities are tightly coupled to new applications of biotechnology in the pharmaceutical and medical area and in addition form the basis for the development of new methods, for example in genome research. Therefore, besides the quality of the knowledge generated in public and private research centres and universities, a crucial characteristic of the research base is its openness and accessibility for biotechnology firms. The importance of information networks and the propensity for co-operation within the research community will be addressed separately as know-how transfer.

As important indicators for the current situation and future perspectives of the Hungarian research base, the amount of financial resources devoted to research, the international patent activities of Hungary and the technological standard of firms and institutes are considered in the following.

Figures 4.2-1 and 4.2-2 show the dramatic decline of gross domestic expenditure on R&D (GERD) calculated on purchasing power parity: business expenditure especially has deteriorated sharply, consequently, the percentage of government funding has increased, but also declined in real terms (see chapter 3.2.1). In addition, from 1994 to 1995 there has been a budget cut of about 50 % (Inzelt 1995b). In total, in 1994, Hungary devoted 1 % of GDP to R&D (Inzelt 1995b), compared to Japan with 2.94 % of GDP, the US with 2.66 %, Germany with 2.33 % and the UK with 2.20 % (OECD 1996). The Hungarian ratio has reached a level between Spain and Italy (at least until 1993), however, it is clearly below EU average (figure 4.2-1).

Figure 4.2-1:   Gross Domestic Expenditure on R&D as a Percentage of GDP

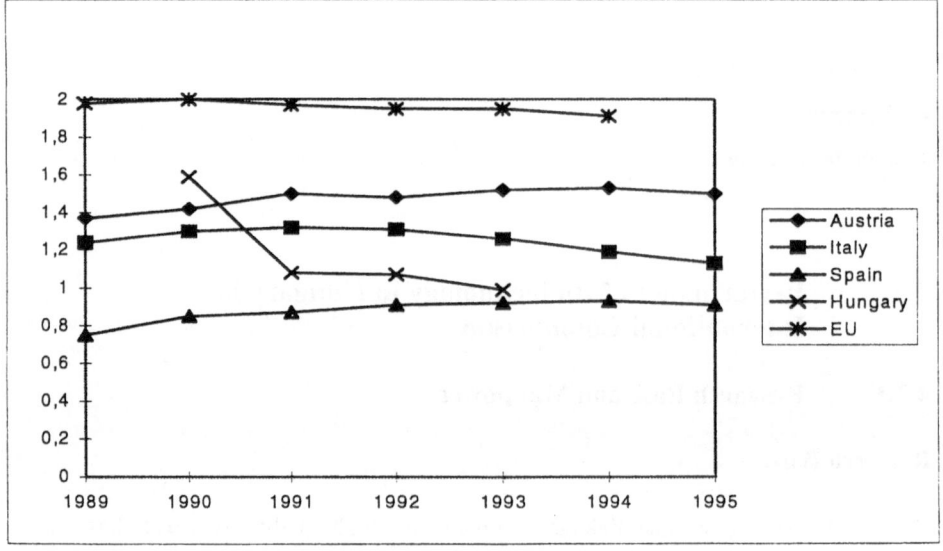

Besides the share of GDP the absolute amount of financial resources is very important to evaluate whether the critical mass for research in the specific technology field is achieved in the country.

Figure 4.2-2:   Hungarian Gross Domestic Expenditure on R&D in Million US$
                 (Purchasing Power Parities)

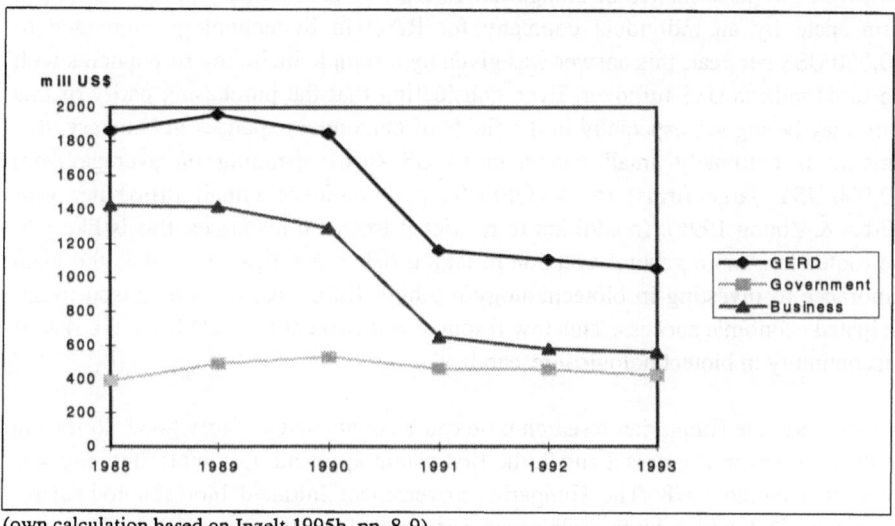

(own calculation based on Inzelt 1995b, pp. 8-9)

More details on state and business expenditure on R&D will be analysed. Still, biotechnology has a high priority, since its share among public funding of applied research through OMFB over the years 1991 to 1996 amounted to 11.5 %, which
means 100 million US$ in five years, or 20 million US$ per year (calculated at an
average exchange rate of 126 HUF/1 US$). Compared to some of the leading coun-
tries in biotechnology, like the US with 4 billion US$ in 1993 and 4.3 US$ in 1994
for the funding of biotechnology, Germany with 1.35 billion DM per year total
public funds devoted to biotechnology, and the UK 0,45 billion £ channelled
through the research councils towards biotechnology, the Hungarian share is very
limited. While the international trend is to increase funds for biotechnology, there
are no signs to be found that resources in Hungary will be increased (chapter 3.2.1).

Internationally, the importance of basic research is broadly acknowledged: the
Japanese policy not to invest heavily in basic research has been a competitive disadvantage for which strategic co-operations with US institutes could not compensate successfully (see chapter 2). Recently, Great Britain has increased the share in
basic research of total civilian R&D expenditure from 35.1 % in 1985-86 to 54.9 %
in 1994-95 (Nature 23 May 1996). On the contrary, in Hungary there seem to be
severe resource cutbacks, especially in some institutes of the Hungarian Academy
of Sciences which formerly performed basic research as a major player. There is no
statistical evidence whether this is part of a more general strategy to discriminate
against basic research, but may have severe implications for the future of biotechnology in the country.

In the business sector, the picture does not change much. In contrast to small US firms which are characterised by a high research intensity with up to 90 % of their turnover, the performance of Hungarian enterprises is extremely poor. The largest sum spent by an individual company for R&D in biotechnology amounted to 80,000 US$ per year, this answer was given by a sample including respondents with up to 41 million US$ turnover. Even considering that the purchasing parity of this sum may be higher, especially in the fields of personnel expenses in Hungary, this amount is extremely small compared to US firms spending on average from 52,000 US$ (large firms) to 145,000 US$ per employee (small firms) per year (Ernst & Young 1996). In addition to restricted financial resources, this is likely to be rooted in a more general aversion to taking risk, since firms ranked as the main hindrance to investing in biotechnology too high R&D cost by comparison to anticipated economic success. This low resource endowment to R&D bears the risk of discontinuity in biotechnological research.

Historically, the Hungarian research base can be evaluated as fairly good. Referring to the analyses in chapters 2 and 3, the first patent application in biotechnology was submitted in the 1978. The Hungarian government initiated biotechnological research in 1982, with a large-scale programme introduced finally in 1984. Compared to the US, where a landmark is constituted by the foundation of the first biotechnological firm in 1976 and Germany and Japan starting public support of biotechnology in the early 1980s, Hungary is no late-comer but has a considerable time span of experience in biotechnological research.

Among other eastern European countries, through the years 1985 to 1989 Hungary has in total almost as many patents granted in the US as the USSR, falling behind slightly after 1989, but still far ahead of all other former CMEA countries (Patent and Trademark Office 1992). The strong position of institutes could be maintained through 1993 with only little decline, especially by institutes in biotechnology-relevant areas such as in the organic/inorganic chemicals and pharmaceutical sectors (Inzelt 1995b). On the other hand, domestic patent applications in Hungary show decline as well as of institutional as individual applicants (see chapter 3.3.5.3), while at the same time the applications for patents on behalf of foreigners in Hungary has considerably increased. In order to obtain a more differentiated picture for biotechnology, additional specific patent analyses would be needed.

An internationally comparable assessment of the technological strengths of Hungarian research institutions is very difficult because there are a great variety of institutions which differ considerably in their technological capabilities. Most institutions consider themselves as average with respect to international standards. The audit has shown that in principle all established biotechnological methods and techniques are also used in at least some Hungarian research institutions. However, some important new developments, especially in biotechnology based on pharmaceutical research like combinational approaches and novel screening methods, are not yet employed. Therefore, the more sceptical self-estimations of future performance by certain bigger and more internationally oriented institutions seems to be more appropriate.

Manpower

In order to keep up with the highly dynamic interdisciplinary development of biotechnology, highly qualified and flexible manpower is necessary, requiring appropriate education systems. In addition, proper payment of scientific work, in other words competitive salaries, are needed so that working in the area of biotechnology stays attractive.

According to the evaluation of Hungarian institutes and firms, the quality of human resources is ranked very high. This positive ranking is supported by output indicators of R&D, such as patents granted abroad (see above) and publications. By contrast, in-house resources are in general estimated low: the current situation is characterised by heavy brain drain of the best scientists to western Europe and the US and internal brain drain of researchers to non scientific jobs in industry. High mobility is especially true for the top level scientists. As the British example shows, unfavourable working conditions for biotechnology scientists can threaten even advanced and leading countries in biotechnology and top scientists are attracted to locations in the US. In contrast to the UK, Hungarian biotechnology has not only to compete with other countries at margin: research institutes often can only afford to pay scientists salaries below living costs. Working conditions are characterised by loss of technical staff, no money to operate and maintain equipment or buy materials, thus effectively impeding researchers' work. Moreover, pessimistic future perspectives hinder especially long-term oriented research which is essential for biotechnology. Almost all interviewed institutes reported decline in scientific staff during the transition period. This audit result is supported by recent statistical evidence on overall R&D personnel in Hungary, which declined about 40.1 % in the business sector and 16.8 % in the public sector (United Nations 1996).

The education system is only affected to a lesser extent by the budget cut backs, so that the education of highly qualified personnel is guaranteed. Anyway, universities formerly did not rank among the institutions in which the highest research expertise was concentrated. This strict division of tasks between research institutes and universities in Hungary is part of the legacy of the socialist organisation of R&D. In advanced market economies, universities perform education and research which makes it possible to benefit from arising synergies.

Summary

In conclusion it can be stated that, both, the Hungarian research base and manpower as key factors for success in biotechnology have a high standard but are severely threatened by lack of financial funds and the consequences thereof. Especially in a highly dynamic science field like biotechnology where knowledge and technology are very quickly outdated, it is not clear to what extent the standard has declined and if the damage is reversible.

4.2.2   Know-how Transfer and Interdisciplinary Networking

Industrial Structure

Industry structure in biotechnology is shaped by a huge share of SMEs and dynamic start-up activities in most countries, especially in the US, UK, Israel and to a lesser degree in Germany. In Japan, SMEs do not play a significant role at all. In Hungary, SMEs amount only to 25 % of biotechnological firms, while the major part of firms has more than 100 employees (70 %)[52]. Two reasons contribute to this high share: large companies now mainly owned by foreigners operate especially in the pharmaceutical sector. In addition, biotechnological units still belong to large complexes as the privatisation and restructuring process has not been fully concluded yet.

Start-up dynamics are estimated to be only of little relevance, since there were neither any data nor indications to be found in interviews with firms and institutes. Instead, a considerable number of firms formerly engaged in biotechnology have moved out of this business.

Figure 4.2-3 represents a rough qualitative ranking (which is not based on exact statistical data) to illustrate the position of biotechnology industry in the analysed countries as described in chapter 2, taking into account their start-up dynamics in numbers and trends as well as the predominant size of companies.

---

[52] Considering the number of firms which refused to respond to the interviews, the amount of 18 covers a huge part of the Hungarian biotechnology industry, compared to Israel with about 150 biotech firms and Germany and the UK with around 400 firms. The answers therefore can be interpreted as representative.

177

Figure 4.2-3:   Industrial Dynamics in Biotechnology

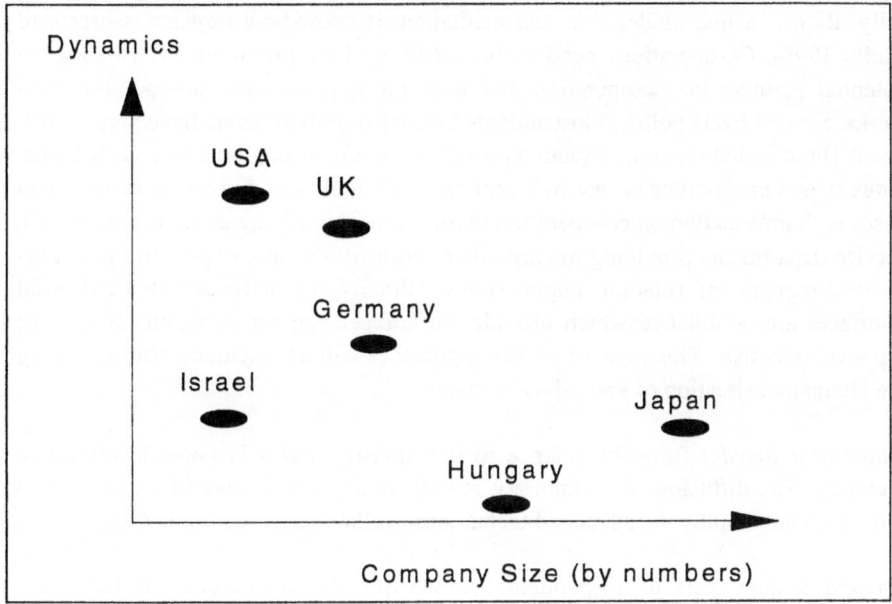

To conclude from the empirical evidence, so far no trend towards a broad basis of dynamic and flexible enterprises can be observed in Hungary. Therefore, the application and market introduction of biotechnological research poses an important problem and so far, has to rely on few players, among them foreign-owned companies.

Know-how Transfer

Know-how transfer aims at the exploitation of research results and the diffusion of new technology: innovation comprises the introduction of research results into the production process. The transfer of know-how is a non-trivial issue because of the basic features of the "commodity" involved. Mostly, the relevant know-how is not freely available but confined to specific individuals or institutes, not only due to organisational and ownership questions but because of its very nature to be not codifiable and disembodied. Know-how transfer is necessary from the major suppliers of new technological knowledge, namely research institutes and universities, to industry as well as among different firms.

There are various mechanisms of know-how transfer between the research base and the industry sector. On the one side the transfer is realised through individuals via spill-out and employment in industry as well as through strategic agreements of firms, such as licensing, acquisition of know-how, interfirm linkages and subcontracting and other forms of networking. On the other side, international experience shows that know-how transfer in biotechnology will not develop automatically. Rather, active moderation and mediation of know-how transfer is practised (Reiss 1996). Co-operations need to be encouraged by providing information on potential partners for co-operation and help for implementing co-operation networks. Several R&D political instruments for realising these goals have been developed. These include specific research programmes supporting joint projects between industry and universities or research centres, workshops and partnering events, data bases with information on co-operation partners, biotechnology parks which include service departments providing among others information on co-operation possibilities and regional information support centres. Finally, it is necessary that industrial interfaces are established which provide the competence for communicating with top level scientists. This issue of policy approaches will be discussed after assessing the Hungarian situation of know-how transfer.

Know-how transfer from the science to the industry sector is underdeveloped in Hungary. The diffusion of technology is only limited, as it was found that firms mostly do not employ the advanced techniques available in research institutes.

Though a huge part of respondents emphasises the importance of R&D co-operations as well as personnel exchange and academic studies, only few firms actually maintain R&D co-operations. Firms name relations with R&D partners only at rank seven out of eleven choices. Consequently, among the preferred sources for innovation conferences and exhibitions rank very high, and informal contacts are among the favourite co-operation modes. Joint research projects which are a very effective method to generate and transfer new technological uncodified know-how are not explored. Informal contacts among specialists from different institutes and companies have shown to be of certain importance even in US high tech regions (often described as the "cafeteria effect"), but have only to be seen as supplementary and cannot compensate for strategic formal co-operation.

In the present situation with shrinking research departments, knowledge transfer through recruitment of academic scholars is not pursued on behalf of firms. Biotechnologists in industry mostly are employed in non-research-related fields where their expertise is not explored.

Another instrument of knowledge transfer from the research sector into industry is the funding of university chairs or researchers at institutes through companies, which can be institutional sponsorship or through project assignments. So far, this international co-operation practice could not be found in the biotechnology sector in Hungary.

On the one hand, most research institutes (with some exceptions) play a rather passive role in search for industrial R&D partners. Notably, Hungarian research institutes are well integrated into the international scientific community through joint projects, conferences and personnel exchange, thereby having access to recent research results on an international level.

On the other hand, some firms were not enthusiastic about co-operation with research institutes or other firms, but this differs considerably with the size of the firm. On the contrary, US experience, for example, points out the growing importance, especially for small firms of contract and collaborative research with institutes and other firms (Ernst & Young 1994). As mentioned earlier, this reflects the need to compete in a highly dynamic and specialised environment in which not all relevant expertise can be available in-house. Nevertheless, in-house R&D and research co-operation are important complementaries. The reluctance to co-operate might be an attitude or financial problem, or both. In Hungary, former experiences in the socialist era have to be considered when analysing co-operation: though there have been collaborations between companies and institutes there were few innovative outcomes. After the break-down of former relations, networks of new market-economy-type between institutes and companies have not yet emerged as substitutes.

Larger firms, some of which are subsidiaries of multinationals, display a different attitude towards co-operation not only with research institutes but also with other companies. Larger companies engage more frequently in collaborations with R&D institutes than smaller firms. Considering the possible mechanisms for technology transfer on behalf of the firm (see above), in the Hungarian case, contacts with other firms, namely suppliers, customers and competitors through strategic agreements are not favoured. While client relationships are more important for smaller firms as well, larger firms again explore the advantages of linkages with competitors and suppliers to a much higher degree and make use of licence agreements. Co-operation with foreign firms still has the lowest priority among forms of R&D co-operation. In contrast, pharmaceutical firms especially in the US, devote huge sums to finance research consortiums or purchase relevant know-how internationally. The benefits of co-operation among firms and with universities and research institutes in the form of strategic alliances or networks has been pointed out by many scientific scholars (e.g. Grabher 1993). Because of the specialities of biotechnology, i.e. its science linkage, dynamic development and the huge investments needed, economic success is believed to be almost impossible (e.g. Powell and Brantley 1992).

To summarise, small firms which mainly are Hungarian owned do not explore enough the modes of know-how transfer and in consequence the technological expertise developed in research institutes is not transferred into marketable innovations. On the other hand, it is not clear and often doubted by interviewees whether foreign firms are willing to absorb and apply the full range of Hungarian expertise to the benefit of the Hungarian economy and thus play an active role in the Hungarian innovation system. In theory, there would be a market for research institutes, as e.g. pharmaceutical firms in Hungary do not perform basic research in biotechnology and applied research only to a slightly higher degree (see chapter 3).

The fact that smaller enterprises participate less in the knowledge transfer is with some exceptions an internationally observable phenomenon. This is related quite often to the size disadvantages which small firms have in comparison to large corporations.

Furthermore, this result is not surprising when the Hungarian practices of public support of know-how transfer is compared with international experience. There are not only different approaches to how firms organise the know-how transfer, but also different public policy approaches to fostering the know-how transfer between the science and industry sector and within the sectors.

Diffusion and application of research findings in new processes and products is to a lesser extent difficult in an environment which is characterised by a high rate of newly founded companies, high spin-off rates out of research institutes and larger companies under favourable surviving conditions, which have been important success factors in the US and UK biotechnology industry. Nevertheless, public policy in most countries has aimed at facilitating the transfer activities to fully exploit the achievements of their national research base.

Technology transfer policy has a long international tradition since the 1970s. Measures are directed towards the diffusion and easy accessibility of research results and new techniques through an effective information infrastructure. A major part of measures is addressed to SMEs and provision or subsidy of services. An increasing trend has been towards regional initiatives (Clement et al. 1995).

The elements of an information infrastructure are knowledge centres at universities and a network of transfer institutes offering services to industry. For example, in Germany, the Fraunhofer Society is a network of specialised institutes with relevance to industry needs, in Baden-Württemberg, the Steinbeis network of transfer institutes offering consulting services to enterprises is attached to polytechnics. In France, similar tasks are performed by national institutes. Additional services are the organisation of conferences, seminars, exhibitions, data bank services etc. Technology parks are created to provide a favourable environment for new companies. Government programmes subsidise co-operation between firms and industrial research, especially in the pre-competitive field. In addition, special programmes with own budget are launched, like the Bio-Regio-Programme in Germany. Postgraduate education with relevance to industrial needs is sponsored. In the UK, a foresight programme has been introduced to reduce uncertainty in research, to name only a few examples.

In western Europe, intermediary institutions play an important role, among them universities and public research institutes as well as chambers of commerce, have a specific task in the transfer of knowledge. The know-how transfer depends to a large degree on the quality of these intermediary institutions. For Hungary, both firms and research institutes rank government as their most important partner. Furthermore, they also rely to a large extent on state measures to solve their current problems. Considering the importance of the state within the restructuring period and the lack of financial resources of all actors, the state clearly plays an important role, but innovation is in its essence not a state responsibility but depends on the creativity of entrepreneurial researchers, engineers and market specialists. Self-organised institutions like associations mostly do not have the expertise or the primary aim of supporting their members, as was expressed during the interviews.

Interdisciplinary Networking

Due to the multidisciplinarity and modular structure of biotechnology, different scientific disciplines and technologies need to be integrated. This means that there is not a one-way first-choice route of development; rather, it is necessary to have knowledge from different disciplines like life sciences, chemistry, physics, mathematics, informatics, engineering, material science available and integrate it into the combinations needed for certain biotechnological innovations. In order to provide these networks, communication and information are important requirements. On the firms' side it also implies that in many cases the necessary knowledge cannot be provided by a single company. Rather strategic research co-operations will be necessary.

As shown already in the context of know-how transfer, multidisciplinary networking is underdeveloped in Hungary, considering the reservations against co-operation in others than informal ways. Though Hungarian firms and institutes are members of several associations, they have not appeared to exert important impact on innovation and interdisciplinary problem-solving. On the contrary, the interdisciplinarity and the many areas of activity in biotechnology are factors which are difficult for smaller actors to handle when searching for partners.

4.2.3 Market Orientation

The long time-to-market periods in many important areas of biotechnology require that during a rather long innovation process the market orientation must be developed. This means that market entrance conditions, market structures and dynamics must be considered even at rather early stages in order to combine the long-term and risky R&D activities with the expected market requirements and adjust them accordingly.

For new biotechnology firms this means that it is necessary to have the economic expertise needed for developing these market issues available in-house. In addition, economic knowledge is also necessary for internal management. Therefore, it is not enough to base the foundation of a new company just on a promising scientific result or method, rather it is necessary to combine the scientific and technological issues with a strong market vision from the early beginnings. If the firm's internal resources are not sufficient to provide this knowledge, it will be necessary to draw on external information related to economic issues, which could be for example provided by regional biotechnology information centres (see above).

Hungarian institutes' main aim is applied and to a slightly lower degree basic research, they also perform experimental development, pilot standards and are engaged in market introduction and market diffusion. Still, this does not mean market orientation in itself, but results from former financing strategies when institutes had to fund basic research through marketable activities. Further, this means that they cannot concentrate on their core competencies and are too small to achieve economies of scale in production. Thus, the division of labour between institutes and market partners is not yet clear but needs repositioning.

Moreover, market-orientation for research institutes should rather examine the market for selling their research results instead of a final product. Their "product" is an intermediary product in the innovation process. Primary concern of institutes and universities is not market-orientation in the sense of how to find application for their research results but they regard scientific publications as their main output. But contract research in highly sensitive and competitive fields like biotechnology with firms often inhibits publishing. Alike, application for patents postpone publishing for some time, though creating long-term income opportunities. This means, that the incentive mechanisms - the evaluation of scientists and therefore future career opportunities are to a large extent based on publications - prevent market orientation. It should be pointed out that the rather small orientation towards market and demand is not a specific problem of Hungarian institutes, but can be observed internationally.

The main domain of firms is market diffusion; only some firms are engaged in applied research, experimental development, pilot standards and market introduction. Referring to their goals for innovation, market-orientation is not concretised, rather process and product innovation in general and following technological trends are named. The own evaluation of marketing position and future aspirations is quite contradictory, there is no strategy to realise market hopes. The first priority of firms is to increase market share or substitute phased-out products. Furthermore, in the future, market shares abroad including EU and other industrialised countries as well as former CMEA countries should not only be defended but increased. In contrast to this optimistic view, on the other hand, marketing and sales capabilities are ranked as very bad, management in general only average, and lack of knowledge about markets is seen as the third important factor among the obstacles to success.

To summarise, the market objectives of firms are mostly not operationalised. A "market vision" is no guiding impulse for R&D in biotechnology. Strategy instruments such as market research are only applied as rare exceptions. There are no conceptual frameworks to analyse and respond to the challenges of the market environment. This is especially true because of the domination of large foreign corporations. Taking into account the areas in which Hungarian firms are active, like the pharmaceutical industry, business is global and strategic management essential.

Like in the leading countries, especially the US and the UK as well as Israel and to a lesser extent Germany, the health care sector which is showing a highly dynamic development and offering marketable applications of biotechnology plays an important role in Hungary - with the restrictions mentioned above. Considering the historical strengths of Hungary in agricultural research, the marked specialisation is understandable.

4.2.4   Financing

As a consequence of the long time-to-market periods firms need to find appropriate financing strategies. These have also to take into account the rather high capital demand and technical riskiness of many biotechnological innovations. According to international experience, venture capital alone is not the key for private financing of biotechnology. Venture capitalists are calculating with a rather narrow time period of approximately four years. Therefore, exit options are a key requirement for venture capital financing. The American experience shows that public offerings and equity from strategic alliances are important ways of exit strategies combined with venture financing. The shortage of private venture capital can be overcome to some extent by state schemes to support young technology firms, like for instance in Germany.

Financing basic research is an unequally harder task compared to raising capital for market introduction. Groups of big firms may overcome this problem internally or through strategic alliances while this is almost impossible for small innovative companies. As international evidence shows, basic research is performed mostly in universities and research institutes for which state or federal governments provide the means. Besides this, areas of public support are co-operation projects to enhance the know-how transfer between academic institutions and industry partners. Programmes targeted towards specific biotechnologies have not been shown to have high measurable impact on the competitive position of industries.

In Hungary, financing is considered overall as the crucial factor impeding innovation (see chapter 3.3.5.1). Large multinational companies can raise money outside Hungary, but all other companies face severe financing restrictions. Financing perspectives are poor on loan as well as on equity basis. As many loans have been given to now insolvent enterprises, the financing for prospering firms is even scarcer. The Hungarian banking system is underdeveloped and many banks are close to ruin, seeking international investors. (Business Central Europe 1996a). As banks are no longer perpetuated as in other central and eastern European Countries, this could be the basis for restructuring and a competitive system in the future.

Venture capital is absent. In central Europe, institutional investors like insurance companies, which for instance account for 50 % of the stock market investment in the UK, invest in safe government papers or even hold bank deposits. (Business Central Europe 1996b) Especially in Hungary, short-term government papers are attractive because of high inflation and accordingly high interest rates. For the corporate finance sector, the unfavourable situation of unfinanciable credit rates will not change until government can control public debt and inflation. The only possibility to raise financial funds is to find foreign investors, which, however, does not mean that they will supply resources for domestic R&D.

For American firms, the stock market is an important capital source for initial financing or follow-on financing. Calculated on employee numbers, half of the US biotechnology companies are public companies (Ernst & Young 1994a). In Europe, especially the listing requirements on the London Stock Exchange are considered to be especially favourable for biotechnological companies (see chapter 3.2.1), though, on the other hand, investors are still less familiar with the risk assessment of biotechnology than their US counterparts. A great impact is believed to result from the EASDAQ (European Association of Securities and Dealers Automated Quotation) in 1996, which constitutes a single European market for financial services (Ernst & Young 1995, 1996). In the Hungarian case, so far the listing of biotechnology companies and raising respective high volumes is not realistic because of a largely dry capital market. Nevertheless, in comparison to the other central European trading places the Budapest Stock Exchange shows the best performance in the recent past, so this could be a future perspective.

Until the stabilisation of the private economy and the establishment of a viable financial system have been achieved, government funds are a very important source for funding, unless firms can earn some money on the market or raise financial means elsewhere. The role of government funds is for example seen in all the countries in which the venture society is not very well developed. This can be seen especially in Germany, where the Ministry for Education, Science, Research and Technology has established several programmes to support the availability of venture capital for small and medium-sized technology enterprises. In order to compensate for the underdeveloped venture society, public venture capital funds have been established, they provide equity and assist young firms through consulting. Furthermore, special programmes targeted towards biotechnology have been launched.

4.2.5   Legal Framework

A prerequisite for international competitiveness is the compliance of national legal frameworks with international standards, which include, for example, safety issues and intellectual property rights. Thereby, not only planing security for research and industry can be provided but also the needed confidence and acceptance among the consumers, which will be conclusive for the market success of new biological products. On a European level the EU guidelines and directives have created the legal framework which is being implemented in the different member states.

Closely related to the financing issue, the fiscal law has to set the right incentives to invest money in research and development, e.g. through tax allowances and similar instruments.

Already, some important elements of a modern legal framework for biotechnology have been introduced in Hungary, especially in the fields of patents and protection of intellectual property rights. Safety standards and consumer protection are not effective so far. The interviewed firms and institutes ranked regulations as average. While in some areas missing regulations can be seen as a short-term advantage to attract research which cannot be carried out elsewhere in Europe, the mid-term perspective requires the establishment of a reliable scheme. Uncertainties must be avoided about the effective implementation of European biotechnology standards, which are necessary for unproblematic approval of Hungarian products in the EU and for closer co-operation.

Among the business laws, the strict Hungarian bankruptcy law of 1992 is a considerable burden on biotech firms in the light of their specific features of long R&D periods and long time-to-market periods until subsistent cash-flow arises. Staying in business is largely dependent on long-term capital availability, the survival index for the US and Europe shows the in limbo situation of many bio-tech firms. In Hungary, creditors may request bankruptcy after 60 days of payment delay, which means a considerable threat for young biotech firms (OECD 1993).

5. Conclusions

In the introduction the question was posed, whether and to which extent biotechnology may, in terms of comparative advantage, become an industry of the future for Hungary which accordingly should be promoted in an appropriate way. The results of the audit indicate that Hungary may expect a good position in *agro-food biotechnology* while other biotechnology-related sectors look less promising. The agro-food biotechnology area could be a strategic one for Hungary. On the other hand it is very doubtful whether biotechnology in the pharmaceutical industry will have strategic importance in Hungary. This does not mean that this sector should be neglected. However, since it is important for success to concentrate the critical innovation mass, government policy has to choose and select target areas and target groups for its supporting measures. In this context the audit provides an additional tool for priority-setting in policy formulation.

Creation and diffusion of new technologies involves a dynamic interaction between firms and their environments. These environments, called "innovation infrastructures", consist of networks of suppliers and customers, support services, financial institutions, as well as the publicly supported science and technology infrastructure, such as universities, public R&D programmes.

The present Technology Audit highlighted some strengths and weaknesses at both micro and macro levels. One of the most critical problems is the country's weak knowledge distribution power and consequently the inadequate utilisation of available knowledge by firms. The reasons for isolation and underdeveloped co-operation among potential partners of the innovation process were discussed in detail. As a consequence commercialisation of the results of earlier research and experimental development is very slow.

In principle we share the opinion of many of our respondents that *biotechnology, or more specifically, certain sub-areas of biotechnology like agro-food applications, could be a strategic issue for Hungary*. However, this does not mean that everything related to biotechnology has strategic importance and all R&D costs have to be financed by public sources. The role of the government in this context is on two different levels: the first one comprises the design of an adequate framework, especially through legislation, so that environmental conditions are favourable for research, creativity, entrepreneurial spirit, and financing. Besides setting the right incentive scheme, influencing the actual environmental conditions is only possible in the long run and dependent on many other factors which partly are not within the government's sphere of influence. Furthermore, the government has responsibility in those areas where market failure occurs. In the case of biotechnology, this is true for basic research and the education system. Both constitute important inputs into the innovation process, but their private provision is unrealistic. Among the success factors for biotechnology state-funded basic research has ranked very high, as shown by the US and UK as well Japanese experiences. Accordingly, government is the sole or an important player with respect to the following success factors: research base, manpower, legal framework.

Secondly, government has to intervene in the short run, to compensate for the unfavourable conditions for biotechnology. This is the government policy most often pursued internationally. It is a logical consequence to explore and appropriate the research results. Success factors which need support measures are especially know-how transfer and interdisciplinary networking and financing. In this context, international experience with different policy instruments, e.g. practices of direct and indirect funding in Germany, could be used when developing respective policy measures.

In general, many problems hampering the development of biotechnology originate in the global economic situation like budget cuts, high interest rates due to high inflation, underdeveloped purchasing power, not fully effective company restructuring and others. The unstable Hungarian market and changeable economic conditions are not favourable for strategy formulation and strategic analyses are usually not well founded. In the following policy implications and recommendations based on the results of this audit will be summarised, taking account of these difficulties. These will touch on three levels: general issues; framework conditions for biotechnology; research institutions, business and innovation networks.

General Issues

- In general terms, the privatisation process has to be finished, re-deployment of organisations by the state has to be settled soon. Imperfect legal regulation should be improved and rules made clear.

- Concerning the future of biotechnology firms, we have to emphasise that the financing system of R&D and innovation are key issues for the Hungarian market economy model. Two large public funders (OMFB and OTKA) can hardly fulfil their tasks because the shortage of money is a great obstacle to efficient working. There are no clear priorities if we investigate the declared aims from the distributing side of state (funds) financial sources. Industrial research carried out and financed by businesses is on a small scale. In-house expenditure on R&D including biotechnology is very limited. Spending on in-house research is important, even if firms do not like to produce new scientific results. According to international experiences, if a firm does not have any R&D capability it is hardly able to adopt (acquire) new technologies. Even follower companies need some R&D capacity if they wish to adapt effectively. Summing up firms' opinions we do not have too many illusions regarding business-financed projects. Some of them are waiting for new results and would be glad to commercialise right away, but they think government has to cover R&D expenditure at institutes. Except for the traditionally knowledge-based pharmaceutical industry, business has to learn that R&D output is not available without costs for them. If firms would like to remain players on the biotechnology stage they should consider R&D partners more seriously. If they need new R&D results to improve their competitiveness they have to invest in them.

- One of the current strengths of the Hungarian biotechnology sector is the good quality of human resources. But at the same time there are not enough well-educated specialists. Both government and business have the duty to create more possibilities (or at least to maintain the existing ones) for *education and training of young specialists*. If no one invests in education and post-graduate courses any more, the quantity of qualified people will become one of the greatest impeding factors of further development.

Framework Conditions for Biotechnology

- It has advantages and also disadvantages that a specific *biotechnology law* has not yet been enacted. While such a legislation might create obstacles for performing certain biotechnology activities which are under legal constraints in other countries, a missing special biotechnology framework on the other hand leads to an uncertain and changeable situation with negative impacts on planning security in firms. In addition, if Hungary wants to become an international player in biotechnology, it will be necessary to adopt international legal standards. This applies, for example, to the enactment of the long-discussed biotechnology law, and also to the international convention on animals. In addition, revision of slow and costly approval procedures in the pharmaceutical and in some cases in the food industry is needed.

- Until now there has been no public debate about biotechnology in Hungary and there is no lobby against biotechnology, which can be considered as a strength and opportunity of this sector. The country can avoid the emergence of a negative public attitude towards biotechnology if public knowledge is increased in time so that the public can decide, on a well-informed basis, about e.g. accepting biotechnology related products.

- For both policy-makers and industry an *up-to-date data bank* would be helpful as an information base in order to develop and implement policy measures and firm strategies more efficiently. This follow-up audit invested a lot to identify biotechnology organisations. Its results offer a good starting point to develop an up-to-date register and data bank which could complement the register on agriculture-related institutes which has been prepared recently by the Ministry of Agriculture.

- Insufficient *demand for biotechnology, know-how, products and processes* is a burdensome factor for institutes. Relations between the science and industry sectors are weak but neither are co-operations within the industry sector strong. If the lack of co-operation persists, achieving a competitive position in biotechnology will be impossible given highly flexible and fast acting international players. Furthermore, the lack of co-operation may lead to the loss of potential partners in the future. Government can encourage university/industry linkages, offer support to establish technology transfer organisations and create new possibilities *to join European programmes.*

Research Institutes, Business and Innovation Networks

- *Benefit of intellectual assets* may become a source for further development and to this end more attention should be paid to patents. The attitude of inventors towards patent applications has to be changed. This is not only a task for government, but for R&D organisations, too. Institutes need to strengthen their patent activities and marketing, and make a *selection in their research portfolio,* taking into account criteria related to the scientific value of projects (mainly a basic research organisations) and to value and demand (mainly applied research and experimental development organisations). Institutes are waiting for firms as partners to finance their research projects and experimental development. However, their "supply" has to meet business demand.

- As the SWOT analysis highlighted, firms and institutes have to *invest more in marketing* and try to *develop links* with each other and with the international community, too. They have to broaden their information sources, concentrating more on business-type information. Institutes have to pay more attention to business demand when offering R&D services. Agricultural institutes are usually ready to meet firms' demands. They are more client-oriented than other research organisations, but they also have to face the problem of a lack of demand. Some basic research organisations e.g. university departments, declared they are ready to continue applied research or experimental development on a contract basis to earn money to develop their laboratories and to receive normal wages.

- In many cases R&D organisations need a *new style of management.* Management of scientific organisations is a problem, not only in transition economies, but all over the world. At the end of the 20th century a brilliant scientist is not automatically the appropriate leader for a research organisation. The problem goes deeper in transition economies than in advanced market economies because, market-oriented practices such as fund raising were out of the question until the late years of socialism.

This Technology Audit helps to clarify the nature of the "Hungarian genius" as it is manifested in biotechnology science and technology and its potential contributions to the economic and social development of the nation, which now finds itself, following major political change, involved in intense global economic competition. This strategic effort could also be useful to prospective foreign investors and partners, because it would reduce the often prohibitive "search cost" for them. In addition, this Biotechnology Audit may contribute to a development which had been initiated by the OECD pilot Technology Audit namely that technology audit will become a part of Hungarian policy thinking. Discussions with participating representatives of universities, institutes, industry and policy-makers during the workshop at the end of this Biotechnology Audit project confirme this notion in a sense that not only the main findings of this audit were supported but also that the process of the audit itself was rated as an informative and helpful experience.

This audit also provided a lot of methodological results and perceptions. It may improve the efficiency of the dissemination of technology audit knowledge if domestic institutes can participate in every phase of the work. Co-operation between two institutes was a significant element of the follow-up audit and at the end of the project we could evaluate it as beneficial. The core part of the method is relevant for investigation of other sectors. It is an important task to make technology audits which are conducted in different sectors at the same time as comparable as possible. The general conceptual elements can serve as a Technology Audit Guideline for central and east European countries. It may disseminate the technology audit experiences in other central and east European countries, either in training seminars or with audit actions.

6. References

Aigner, H.; Barisitz, St.; Fink, G. (1993): Unternehmensbewertung in Osteuropa. Methoden und Fallstudien. Wiesbaden.

Albach, H.; Witt, P. (Hg.) (1993): Transformationsprozesse in ehemals volkseigenen Betrieben. Stuttgart.

Antalóczy, K. (1996): Nagyvállalatok a gyógyszeriparban (Large firms in the pharmaceutical industry), Pénzügykutató Rt, mimeo.

Apolte, Th. (1992): Politische Ökonomie der Systemtransformation. Gruppeninteressen und Interessenkonflikte im Transformationsprozeß. Hamburg.

Arbeitsgemeinschaft deutscher wirtschaftswissenschaftlicher Forschungsinstitute (1990): Fragen zur Reform der DDR - Wirtschaft. Tagungsband zur Sondertagung am 12.02.1994 in Bonn. Beihefte der Konjunkturpolitik, Zeitschrift für angewandte Wirtschaftsforschung, Heft 37. Bonn.

Arbeitsgemeinschaft deutscher wirtschaftswissenschaftlicher Forschungsinstitute (1993): Wirtschaftsreformen in Mittel- und Osteuropa. Tagungsband zur Jahrestagung am 14. und 15. Mai 1992 in Bonn. Beihefte der Konjunkturpolitik, Zeitschrift für angewandte Wirtschaftsforschung, Heft 40. Berlin.

Auvinen, H. (1994): Premises and Barriers to Commercialisation of Russian IT, Paper presented at the Six Countries Programme Conference on Innovation 'Research Co-operation with Countries in Transition' Vienna, Dec. 1-2.

Aydalot, P. (Ed.) (1986): Milieux Innovateurs en Europe, Paris.

Balazs, K. (1994): Transition Crisis in the Hungarian R&D Sector, in: in: Economic Systems, Vol. 8, p. 281 - 306.

Bartol, K.; Martin, D. (1994): Management, 2nd ed., McGraw-Hill: New York.

Batstone, S.; Weszhead, P. (1996): The development of science parks and new-technology based firms in Russia, in B/F p. 72-93.

BBSRC (1995): Corporate Plan 1995-1999. Swindon.

Becher, G.; Kuhlmann, S. (eds.) (1995): Evaluation of Technology Policy Programmes in Germany, Kluwer Academic Publishers, Dordrecht, Boston, London.

Bentley, R. (1992): Research and Technology in the Former German Democratic Republic. Boulder, San Francisco, Oxford.

Besters, H. (Hg.) (1992): Vitalisierung der ostdeutschen Wirtschaft. Öffentlicher Sektor und private Wirtschaft. Baden-Baden.

BIO (1996): The US Biotechnology Industry Organisation: Facts and Figures 1994/95 Edition. Available through internet.

BioCommerce Data Ltd. (1995): The U. K. Biotechnology Handbook '95. London.

Borsos, J. (1994): Foreign Companies in Estonia - FDI Policy Implications, Paper presented at the Six Countries Programme Conference on Innovation 'Research Co-operation with Countries in Transition' Vienna, Dec. 1-2.

Bozeman, B.; Melkers, J. (ed.) (1993): Evaluating R&D Impacts: Methods and Practice, Bosten/Dordrecht/London.

Braun, T.; Schubert, A. (1997): Indicators of Research Output in the Sciences from 5 Central European Countries 1990 - 1994, in: Scientometrics, forthcoming.

Brezinski, H.; Fritsch, M. (1996): The Economic Impact of New Firms in Post-Socialistic Countries - Bottom-up Transformation in Eastern Europe, Edward Elgar, Cheltenham, Brookfield.

Brower, V. (1996): Alliance seeks to partner US and Israeli biotech companies. GEN 16/16, September 15, 1.

Bullock, W.; Dibner, M. (1994): The changing dynamics of strategic alliances between US biotechnology firms and Japanese corporations and universities. Tibtech 12, 397-400.

Bullock, W.; Dibner, M. (1995): The state of the US biotechnology industry. Tibtech 13, 463-467.

Bundesministerium für Wirtschaft (Ed.) (1996): Wirtschaftslage und Reformprozesse in Mittel- und Osteuropa, Bonn.

Business Central Europe (1996a): Hope and Despair. July/August, 53.

Business Central Europe (1996b): Institutional Investors Dormant, 57-58.

Cabinet Office, Office of Public Service and Science, Office of Science and Technology (1995): Forward Look of Government-funded Science, Engineering and Technology. Volume 2. London: HMSO.

Camagni, R. (1991): Local "milieu", uncertainty and innovation networks: towards a dynamic theory of economic space. In: Camagni, R. (Ed.): Innovation Networks. London, New York.

Carlberg, M. (1994): Makroökonomische Szenarien für das vereinigte Deutschland. Heidelberg.

Clement, K. et al (1995): Regional Policy and Technology Transfer: A Cross-National Perspective, London: HMSO.

Cohen, S. A. et al. (1973): Construction of biologically functional bacterial plasmids in vitro. Proc. Natl. Acad. Sci. USA 70, 3240-3244.

CVCP (1995): Research in universities, A Briefing by the Committee of Vice-Chancellors and Principals of Universities of the United Kingdom. Briefing Note, 1-6.

DesForges, C. (1992): Recent Initiatives in the Commercialisation of Academic Research in Europe - Invention to Innovation, Paper presented at the Conference on Perspectives in Technology Transfer organised by the German Association for the Support of Technology and Innovation, June 1992, Bielefeld.

Dickson, D. (1993): Britain urged to lift barriers to investment in biotechnology. Nature 361, 572.

Dickson, D. (1995a): Millions of Pounds. Confirmed lost to universities over funding switch. Nature 377, 664.

Dickson, D. (1995b): UK goes for generic properties in bid to boost science/industry links. Nature 375, 265.

Dittrich, E.J.; Haferkemper, M.; Schmidt, G.; Stojanov, Ch. (Hg.) (1992): Der Wandel industrieller Beziehungen in Osteuropa. Frankfurt am Main.

Dolgopiatova, T. (1996): The Transitional Model of the Behaviour of Russian Industrial Enterprises (on the basis of regular surveys during 1991-1995), IIASA Working Papers 96-57, Laxenburg.

Dudits, D.; Heszky, L. (1990): Növénybiotechnológia (Plant biotechnology), Mezõgazdasági Kiadó, Budapest.

Dyker, D. (1994): Technology Policy and the Productivity Crisis in Eastern Europe and the Former Soviet Union, in: Economic Systems, Vol. 8, p. 71-85.

Ernst & Young (1994a): Biotech 95: Reform, Restructure, Renewal. Palo Alto.

Ernst & Young (1994b): European Biotech 95: Gathering Momentum. Brussels.

Ernst & Young (1996): European Biotech 96 - Volatility and Value. London.

Europäische Kommission Generaldirektion Wirtschaft und Finanzen (1996): Wirtschaftliche Lage und Wirtschaftsreformen in Mittel- und Osteuropa, Beiheft C Wirtschaftsreformen, Nr. 2 - Juli.

European Commission (1995): The European Handbook of Management Consultancy, Dublin.

Faust, K. et al (1995): Der Wirtschafts- und Forschungsstandort Baden-Württemberg: Potentiale und Perspektiven, ifo Studien zur Strukturforschung; 19/I and II, München.

FCCSET (1992): Biotechnology for the 21st Century. A Report by the FCCSET Committee on Life Sciences and Health. Washington.

FhG ISI (1995): Wirkungsanalyse zum Programm "Förderung der Biotechnologie in der Wirtschaft". Karlsruhe.

Fleissner, P. (ed.) (1994): The Transformation of Slovakia. The Dynamics of Her Economy, Environment and Demography. Hamburg.

Forschungsinstitut für anwendungsorientierte Wissensverarbeitung (FAW) (1993): Wirtschaftliche und politische Öffnung zum Osten.

Freeman, C. (1987): Technology Policy and Economic Performance: Lessons from Japan, Pinter London.

Freemann, C. (1982): The economics of industrial innovation. 2nd copy, London.

Frenkel, A.; Reiss, T.; Maital, S.; Koschatzky, K.; Grupp, H. (1993): Technometric evaluation and technology policy: the case of biodiagnostic kits in Israel. Research Policy 23, 281-292.

Frigyesi, Veronika (1990a): A biotechnológiai innováció hazai lehetõségei. A Mariklón sikerei és kudarcai. (Opportunities in Hungary for biotechnological innovations. Case study on Meriklón.) HAS Institute of Industrial Economics, mimeo.

Frigyesi, Veronika (1990b): A kutatási és fejlesztési erõforrások hasznosulásának lehetõségei a biotechnológia példáján. (Opportunities of applying R&D sources in the case of biotechnology.) Ipargazdasági Szemle 1, 18-25, Budapest.

Frigyesi, Veronika (1993): A biotechnológia fejlõdésének gazdasági feltételei (Nemzetközi tapasztalatok). (Economic conditions in biotechnological development, international experiences). Dissertation, mimeo.

Fritsch, M.; Bröskamp, A.; Schwirten, C. (1996): Innovationen in der sächsischen Industrie - Erste empirische Ergebnisse, Freiberger Working Papers 96/13, Freiberg.

Fröhlich, Z.; Padjen, J.; Svaljek, S. (1995): Partnerships between the public and private sectors - Croatian case, in: Steiner, M. (Ed.): Regionale Innovation: Durch Technologiepolitik zu neuen Strukturen, Leykam Graz.

Gehrig, G.; Welfe, W. (ed.) (1993): Economies in Transition. A System of Models and Forecasts for Germany and Poland. Heidelberg, New York.

Gehrke, B.; Grupp, H. (1994): Innovationspotential und Hochtechnologie: Technologische Position Deutschlands im internationalen Wettbewerb, zweite, vollständig bearbeitete und erweiterte Auflage, Heidelberg.

Georghiou, L. (1995): Research evaluation in European national science and technology systems, in: Research Evaluation, Vol 5, p. 3 - 10.

Gerhardt, U.; Mochmann, E. (1992): Gesellschaftlicher Umbruch 1945 - 1990. Re-Demokratisierung und Lebensverhältnisse. München.

Gibbons, J. H. (1994): Biotechnology: Opportunity and Challenge. National Biotechnology Summit. Washington. Available through internet.

Glismann, H.; Horn, E.-J.; Stanovnik, P. (1995): Institutional Change in Search of the Market: The Case of Slovenia, Kiel Working Paper No. 706, Kiel.

Grabher, G. (1993): The embedded firm. On the socioeconomics of industrial networks. Routledge, London/New York.

Grupp, H. (Ed.) (1993): Technologien am Beginn des 21. Jahrhundert. Heidelberg.

Grupp, H.; Schmoch, U. (1992): Wissenschaftsbindung der Technik. Heidelberg.

Guerrieri, P. (1994): Technology, Structural Change and Trade Pattern of Eastern Europe, Paper presented at the Six Countries Programme Conference on Innovation 'Research Co-operation with Countries in Transition' Vienna Dec 1-2.

György, Katalin (1994): R&D and Co-operation in the Hungarian Pharmaceutical Industry. Report to the World Bank, mimeo.

György, Katalin; Vincze, János (1993): Privatisation and Innovation in the Pharmaceutical Industry in a Post-Socialist Economy. Rivista Internationale di Scienze Sociali, Vol C, No. 3, 403-417.

Hauer et al. (1993): Der Mittelstand in Transformationsprozeß Ostdeutschlands und Osteuropas. Heidelberg.

Hernisniemi, H.; Hyvärinen, J. (1995): Industrial Strategies for Lithuania and Industrial Strategies for Latvia, ETLA, Helsinki.

Hickel, R.; Priewe, J. (1994): Nach dem Fehlstart. Ökonomische Perspektiven der deutschen Einigung. Frankfurt am Main.

Hilbert, A. (1994): Industrieforschung in den neuen Bundesländern. Ausgangsbedingungen und Reorganisation. Wiesbaden.

Hirschmann, K.; Hirschmann, E.; Bode, O.F. (Hg.) (1993): Weltwirtschaftliche Anpassung und Öffnung der osteuropäischen Reformstaaten. Transformationskosten, Handelsstrategien, ökologische Modernisierung, Konsumentenverhalten, Humankapital.

Hohmeyer, O.; Hüsing, B.; Maßfeller, S.; Reiss, T. (1994): Internationale Regulierung der Gentechnik. Heidelberg.

Holland, D. (1995): Evaluation und Transfer eines "Technology Audit"-Verfahrens für MOEL, Zwischenbericht - Ergebnisse der Begleituntersuchung des OECD-Audits in Ungarn, Fraunhofer-Institut für Systemtechnik und Innovationsforschung (ISI), Karlsruhe, März 1995.

Hutschenreiter, G. (1994): Research Co-operation with Countries in Transition, Paper presented at the Six Countries Programme Conference on Innovation 'Research Co-operation with Countries in Transition' Vienna, Dec. 1-2.

ifo (1994): Stand und Entwicklung der Beschäftigung von Biotechnologen in Deutschland. München.

ifo-Institut für Wirtschaftsforschung (1991): The Role of EC Investment in Promoting R&D Capabilities and Technological Innovation in Eastern European Countries: Scientific Goals and Financial Instruments" Interim Report, Part I, CSFR and Poland, München.

Indruch, R. (1994): Der Übergang zur sozialen Marktwirtschaft in den Ländern des ehemaligen RGW am Beispiel der CSFR und Ungarns. Konstanz.

Institut für Wirtschaftsforschung Halle (IWH) (Hg.) (1994): Wirtschaft im Systemschock. Die schwierige Realität der ostdeutschen Transformation. Essen.

Inzelt, A. (1995a): For a better understanding of the innovation process in Hungary, SPRU, STEEP Discussion Paper No 22, University of Sussex.

Inzelt, A. (1995b): Workshop on the Implementation of OECD Methodologies for the Collection and Compilation of R&D/S&T Statistics in the Partner in Transition Countries and the Russian Federation: Methodological Lessons of the Hungarian Pilot Innovation Survey, 4-5 December, Draft version.

Inzelt, A. (1995c): Review of Recent Developments in Science and Technology in Hungary. OECD Paris.

Inzelt, A. (1996a): The Hungarian Industry, in: Widmaier, B., Potratz, W. (ed.): The Future of Industry in Central and Eastern Europe, Institut Arbeit und Technik.

Inzelt, A. (1996b): Institutional Transfer in a Post-Socialist Country (the Case of the Bay Zoltán Foundation for Applied Research and its Institutes), in: CERNA: Innovation in Eastern Europe and Russia - Technical Modernization, Combinates and Enterprises, April, Ecole des Mines, Paris.

Inzelt, A. (1996c): Institutional and Behavioural Conditions for Innovativeness in Central and East Europe, forthcoming.

JBA (1993): Japan Bioindustry Letter, 10/36, 4-7.

Kaiser, K.-A.; Tamm, A. (1992): Osteuropa auf dem Weg zur Marktwirtschaft. Zehn Fallstudien mit Lösungsansätzen. Wiesbaden.

Keren, M. (1996): A dynamic evolutionary perspective on transformation, in: Brezinski/Fritsch, p. 35-51.

Kinoshita, J. (1993): Is Japan a boon or a burden to US industry's leadership? Science 259, 596-598.

Kline, S.J.; Rosenberg, N. (1986): An Overview of Innovation, in: Rosenberg, N./Landau, R., eds.: The Positive Sum Strategy, Washington D.C.

Klodt, H.; Paqué, H. (1993): Am Tiefpunkt der Transformationskrise: Industrie- und lohnpolitische Weichenstellungen in den jungen Bundesländern. Kieler Diskussionsbeiträge, Nr. 213.

Klodt, Henning (1993): Perspektiven des Ost-West-Handels: Die komparativen Vorteile der mittel- und osteuropäischen Reformländer, in: Die Weltwirtschaft, S. 424 - 440.

König, H.; Steiner, V. (Hg.) (1994): Arbeitsdynamik und Unternehmensentwicklung in Osteuropa. Erfahrungen und Perspektiven des Transformationsprozesses. Beiträge eines Workshops des Zentrums für Europäische Wirtschaftsforschung (ZEW) am 4. und 5. März 1993 in Mannheim.

Kosals, L. (1996): Military R&D Institutes in the Context of Demilitarization in Russia, IIASA Working Papers WP-94-002, Laxenburg.

Kuhlmann, S. (1994): Ideas for Discussion, Technology Audit For Central and Eastern European Countries - Workshop on Assessment Approaches of Competitive Technological Innovation Potentials for Transformation Economies, February, Karlsruhe.

Kuhlmann, S. (1995): Patterns of science and technology policy evaluation in Germany, in: Research Evaluation, Vol 5, p. 23 - 33.

Kuhlmann, S.; Holland, D. (1995a): Evaluation von Technologiepolitik in Deutschland: Konzepte, Anwendungen, Perspektiven, Heidelberg.

Kuhlmann, S.; Holland, D. (1995b): Erfolgsfaktoren der wirtschaftsnahen Forschung, Heidelberg.

Kuhlmann, S.; Reger, G. (1996): Technology-Intensive SMEs: Policies Supporting the Management of Growing Technological Complexity, in: Cannell, W., Dankbaar, B.: Technology Management and Public Policy in the European Union, Oxford, Oxford University Press.

Lipsitz, I. (1996): The Dynamics of Russian Industrial Enterprises' Financial Situation (1992-1994), IIASA Working Paper 96-064, Vienna.

Luft, Ch. (1992): Treuhandreport. Werden, Wachsen und Vergehen einer deutschen Behörde. Berlin, Weimar.

Lundvall, B.-A. (1992): Introduction in Lundvall, B.-A. National Systems of Innovation: Towards a Theory of Innovation and Interactive Learning, Pinter London.

Mayntz, R.; Schimank, U.; Weingart, P. (Ed.) (1995): Transformation mittel- und osteuropäischer Wissenschaftssysteme. Länderberichte. Opladen.

Meyer-Krahmer, F. (1987): Evaluating Innovation Policies: The German Experience, in: Technovation, Vol. 5, p. 317 - 330.

Meyer-Krahmer, F. (1991): Technology Policy Evaluation in Germany, Contribution to the SPRU international Conference on Science and technology Policy Evaluation, October 1991, London.

Mohácsi, Kálmán (1996): The Hungarian Food Industry. In: The Future of Industry in Central and Eastern Europe. Institut Arbeit und Technik.

Muller, E.; Gundrum, U.; Koschatzky, K. (1995): Methodology in Design, Construction and Operation of Regional Technology Frameworks. Needs analysis of the innovation and technology support requirements of firms within a region, Fraunhofer Institute for Systems and Innovation Research, Karlsruhe.

Müller, W. (1993): Die Wirtschaftstransformation in Mittel- und Osteuropa als theoretisches Problem. In: Zapotoczky, K. 1993: Mittel- und Osteuropa: Eine Herausforderung für die Unternehmensberatung. Stuttgart, Berlin, Köln.

Münt, G. (1996): Dynamik von Innovation und Außenhandel: Entwicklung technologischer und wirtschaftlicher Spezialisierungsmuster, Heidelberg.

Nature, Vol 381, 23 May 1996, 262.

Nauwelaers, C.; Reid, A. (1995): Innovative regions? A comparative review of methods of evaluating regional innovation potential, RIDER/European Commission, Louvain-La-Neuve/Luxembourg.

OECD (1991a): OECD Wirtschaftsberichte Tschechische und Slowakische Föderative Republik 1991, Paris.

OECD (1991b): OECD Wirtschaftsberichte Deutschland 1991/1992, Paris.

OECD (1991c): OECD Wirtschaftsberichte Ungarn 1991, Paris.

OECD (1992): OECD Wirtschaftsberichte Polen 1992, Paris.

OECD (1993a): Science, Technology and Innovation Policies Hungary, Paris.

OECD (1993b): OECD Economic Surveys Hungary 1993, Paris.

OECD (1994a): Barriers to Trade with the Economies in Transition. Centre for Co-operation with the Economies in Transition. Paris.

OECD (1994b): Science, Technology and Innovation Policies. Federation of Russia. Volume 1. Evaluation Report. Paris.

OECD (1994c): OECD Wirtschaftsberichte Tschechische Republik und Slowakische Republik1994, Paris.

OECD (1994d): OECD Economic Surveys Poland 1994, Paris.

OECD (1994e): Industry in the Czech and Slovak Republics, Paris.

OECD (1995a): OECD Economic Surveys. Hungary, Paris.

OECD (1995b): Review of Industry and Industrial Policy in Hungary. OECD, Paris.

OECD (1995c): Methodology of Technology Audit, Draft by the OECD Secretariate February 1995.

OECD (1995d): OECD Wirtschaftsberichte Die Russische Föderation 1995, Paris.

OECD (1996): Main Science and Technology Indicators, Paris.

OMFB (1995): Innovációs folyamatok a gazdaságban. (Innovation Processes in the Hungarian Economy.) OMFB, Budapest.

Oppenländer, K. (Hg.) (1992): Aktuelle Probleme beim Übergang von der Plan- zur Marktwirtschaft. ifo - IMEMO-Symposium am 2. und 3. September 1991 in Moskau.

OTA (1991): Biotechnology in a global economy. Washington.

Patent and Trademark Office (1992): Highlights in Patent Activity. U.S. Department of Commerce, October.

Porter, M. (1990): The Competitive Advantage of Nations, The Macmillan Press Ltd. London, Basingstoke.

Powell, W.; Brantley, P. (1992): Competitive Cooperation in Biotechnology: Learning through Networks? In: Nohria, N.; Eccles, R. (eds.): Networks and Organisations: Structure, Form, and Action. Harvard Business School Press, Boston (Massachusetts).

Pradetto, A. (Hg.) (1994): Die Rekonstruktion Ostmitteleuropas. Politik, Wirtschaft und Gesellschaft im Umbruch. Opladen.

PROGNOS AG (Hg.) (1992): Vom Umbruch zum Aufbruch. Stuttgart.

Reiss, T. (1996): Knowledge Transfer in Biotechnology - The Case of Germany. In: Inzelt, A.; Coenen, R (eds.): Knowledge, Technology Transfer and Forecasting, 25-31. Kluwer, Dorderecht. Boston, London.

Reiss, T.; Hüsing, B. (1992): Potentialanalyse für Auftragsforschung in der Biotechnologie. Karlsruhe.

Reiss, T.; Jaeckel, G. (1994): Analyse der baden-württembergischen FuE-Strukturen und Potentiale in der Biotechnologie. Karlsruhe.

Research Evaluation (1995): Special issue on National systems for evaluation of R&D in the European Union, Vol 5, No 1 April.

Roeger, W. (1994): Wage Behaviour and Convergence in Open Economies. Lessons for eastern Germany, in: König/Steiner (1994).

Rossi, P.H.; Freeman, H.E. (1985): Evaluation, A Systematic Approach, 3rd ed., Sage Publications, Beverly Hills et al.

Saage, D. et al (1992): Technology Auditing in Germany, Final Report, Fraunhofer Institute for Systems and Innovation Research, Karlsruhe.

Saage, D.; Hemer, J. (1981): Möglichkeiten für den zukünftigen Ausbau des Produktprogramms der Fa. Inovan-Stroebe GmbH & Co. KG, Final Report, Fraunhofer Institut for Systems and Innovation Research, Karlsruhe.

Schikora, A.; Fiedler, A.; Hein, E. (1992): Politische Ökonomie im Wandel. Marburg.

Schimank, U. (1995): Die Transformation der Forschung in Mittel- und Osteuropa: Gelegenheiten, Ziele und Zwänge, in: Wollmann, H., Wiesenthal, Boänker, F. (Ed.): Transformation sozialistischer Gesellschaften: Am Ende des Anfangs, Leviathan Sonderheft 15/1995 Westdeutscher Verlag, p. 321 - 345.

Schmid, R. et al. (1995): Biotechnology in the Asian-Pacific Region. In: Rehm, H.-J.; Reed, G. (eds.): Biotechnology 12, 369-432. Weinheim.

Schmidt, R. (Hg.) (1993): Zwischenbilanz. Analysen zum Transformationsprozeß der ostdeutschen Industrie. Berlin.

Schmoch, U.; Grupp, H.; Laube, T. (1996): Standortvoraussetzungen und technologische Trends, in: Bundesamt für Konjunkturfragen (ed.): Modernisierung am Technikstandort Schweiz, Vdf Hochschulverlag AG an der ETH Zürich, Zürich.

Schmoch, U.; Strauss, E.; Grupp, H.; Reiss, T. (1993): Indicators of the Scientific Base of European Patents. Karlsruhe.

Schmoch, U.; Strauss, E.; Reiss, T. (1992): Patent law and patent analysis in biotechnology. Biotech Forum Europe 9, 379-384.

Schneider, C. (1994): Research and Development Management: From the Soviet Union to Russia. Heidelberg, New York.

Schneider, C. (1996): Post-socialism and Impact of the Economic Reforms on Industrial Research: A Study of Czech and Slovak Research Institutes in the Electrotechnical Sector, in: CERNA: Innovation in Eastern Europe and Russia - Technical Modernization, Combinates and Enterprises, April, Ecole des Mines, Paris.

Schneider, R. (1992): CIM Planning in Small and Medium-Sized Companies, in: Saage, D. (1992).

Schwitalla, B. (1993): Messung und Erklärung industrieller Innovationsaktivitäten: mit einer empirischen Analyse für die westdeutsche Industrie, Heidelberg.

Segal Quince Wicksteed Limited (1995): Technical Assistance for the Restructuring of Science and Technology in Romania, Working materials, Cambridge.

Shama, A. (1995): Developing and Testing a Theory of Management Transformation from Planned to Market Economy: The Case of Russia, in: Technological Forecasting and Social Change, Vol. 48, 77-100.

Shapira, P.; Paskaleva, K. (1994): After Central Planning: The Restructuring of State Industry in Bulgaria's Bourgas Region, in: European Planning Studies, Vol. 2, p.131 - 157.

Shaw, B. (1996): Networking in the Russian aerospace industry, in: R&D Management, Vol. 26 No 3, 255 - 265.

Siebert, H. (Ed.) (1993): Overcoming the transformation crisis: Lessons for the successor states of the Soviet Union, Mohr Tübingen.

Smallbone, D.; Venesaar, U.; Piasecki, B. (1996): Employment Change and Labour Market Issues in Manufacturing SMEs in Poland and the Baltic States: Some Implications for Policy, paper presented at the 2nd International Conference on SME Development Policy in Transition Economies: Centre for Urban Studies, University of Bristol, 24-25 October 1996.

Somogyi, L. (1993): The Political Economy of the Transition Process in Eastern Europe. Proceedings ot the 13th Arne Ryde Symposium, Rungsted Kyst, 11 and 12 June 1992.

Spencer, V.; Kirk, D. (1994): U.K. boasts Europe's biggest public sector.

Statistisches Bundesamt (1995): Ausgaben für biotechnologische Forschung. Stuttgart.

Stoyanovska, A.; Krastenova, E. (1996): Development of the SMEs in Bulgaria, paper presented at the 2nd International Conference on SME Development Policy in Transition Economies: Centre for Urban Studies, University of Bristol, 24-25 October 1996.

Szántó, B. (1995): Science Policy Dilemma: Justification of Unbound Research or Strategic Support of Technology Policy, prepared for Symposium on "Changing Trends in Science Policy: Theory and Practice, Göteborg, Nov. 23-25, Budapest.

Tournemine, R.; Muller, E. (1996): Transition and Development of Innovation Capacities in Hungary, in Research Management, Vol. 26 No 2, 101-113.

United Nations (1996): Review of Changes in National Policies, priorities and Institutions, and International Cooperation. Economic and Social Council.

United Nations (UN) (1992): Biotechnology and Development. Expanding the Capacity to Produce Food. In: Mary Pat, Williams Silviera (eds.): ATAS, Issue 9, New York.

Valigra, L. (1994): Academic biotech deals offer more promise than product. Science 263, 168-169.

Wagener, H.-J. (ed.) (1993): The Political Economy of Transformation. Heidelberg.

Ward, M. (1994 a): UK biotech will boom over next few years. BioTechnology 12, 230.

Ward, M. (1994 b): Dramatic growth forecast for UK biotechnology firms. Nature 367, 674.

Westphal, A.; Herr, H.; Heine, M.; Busch, U. (1991): Wirtschaftspolitische Konsequenzen der deutschen Vereinigung. Frankfurt am Main, New York.

World Bank (1996): World Development Report 1996. From Plan to Market, Washington.

Zapotoczky, K. (Hg.) (1993): Mittel- und Osteuropa: Eine Herausforderung für die Unternehmensberatung. Stuttgart, Berlin, Köln.

Zeschmann, P. (1993): Ideen für ein wirtschaftliches Überleben Ostdeutschlands. Eine wirtschaftspolitische Konzeption zur Bewältigung des Strukturwandels in den neuen Bundesländern. Thun, Frankfurt am Main.

Related Literature:

Balmer, B.; Sharp, M. (1993): The Battle for biotechnology: Scientific and technological paradigms and the management of biotechnology in Britain in the 1980s. Research Policy 22, 463-478, Elsevier.

Biotechnology (1996): U.S. Senate defining, agency defining, FDA reforms 14, 138.

Boldogkõi, Z. (1995): Biotechnológia: Tudomány a gazdaságban. (Biotechnolgy: Science in the Economy.) Gödöllõi Agrártudományi Egyetem, mimeo.

CIS Survey (1993): OECD/EC Harmonised Questionnaire 1992/93, OECD-EUROSTAT.

Crafts-Lighty, A.; Burak Reed, E.; Gallagher, A.; Da Gama, L.: The UK Biotechnology Handbook '94. BioCommerce Data Ltd. and BioIndustry Association.

Faulkner, W.; Senker, J. (1995): Knowledge Frontiers. Public Sector Research and Industrial Innovation in Biotechnology, Engineering Ceramics and Parallel Computing, Oxford University Press.

Fazekas-Horváth, Z.; Csorba, J. (eds.) (1987): A biotechnológia gazdasági vonatkozásai (Economic aspects of biotechnology) Országos Mûszaki Információs Központ és Könyvtár, Témadokumentációs Kiadványok 151. sz.

Fildes, R.A. (1990): Strategic Challenges in Commercialising Biotechnology. California Management Review Vol. 32 No. 3, 63-72.

Forrest, E.; Martin M. J. (1992): Strategic alliances between large and small research intensive organisations: experiences in the biotechnology industry. R&D Management, Vol. 22, No. 1, 41-53.

Grieve, J.; Fleck V. (1988): Strategies of New Biotechnology Firms. Long Range Planning, Vol. 21, No. 3, 51-58.

Hall, S.S. (1988): Invisible Frontiers. The Race to Synthesise a Human Gene. Sidgwick & Jackson, London.

Hamilton, W.F.; Villa, J.; Dibner, M.D. (1990): Patterns of Strategic Choice in Emerging Firms: Positioning for Innovation in Biotechnology. California Management Review, Vol. 32, No. 3, 73-85.

Howells, J. (1994): Managers and Innovation. Strategies for a biotechnology, Routledge. London and New York.

Inzelt, A.; Coenen, R. (1996): Knowledge, Technology Transfer and Forecasting. Kluwer, Dordercht. Boston, London.

Jones, S. (1992): The biotechnologists and the evolution of biotechnology enterprises in the USA and Europe. Macmillan, London.

McKelvey, M. (1996): Redefining Transfer in Biotechnology and Software: Multiple Creation of Knowledge and Issues of Ownership. In: Inzelt, A.; Coenen, R. (eds.): Knowledge, Technology Transfer and Forecasting, 33-47. Kluwer, Dordercht. Boston, London.

Nelson, R. (1993): National Innovation Systems, a Comparative Analysis. Oxford University Press, New York, Oxford.

Oakey, R.; Faulkner, W.; Cooper, S.; Walsh, V. (1990): New Firms in the Biotechnology Industry: Their Contribution to Innovation and Growth. Pinter Publishers, London.

Orsenigo, L. (1989): The Emergence of Biotechnology. Institutions and Markets in Industrial Innovation. Pinter Publishers, London.

Oslo Manual 1992, OECD, Paris.

Pisano, G.P. (1990): The R&D Boundaries of the Firm: An Empirical Analysis. Administrative Science Quarterly, 35, 153-176.

Pisano, Gary P. (1994): Knowledge, Integration, and the Locus of Learning: an Empirical Analysis of Process Development. Strategic Management Journal, Vol. 15, 85-100.

Rudán, F. (1996): Biotechnológia és biztonság. (Biotechnology and Biosafety). Európa Fórum, 2, Budapest, 127-129.

Sapienza, A.M. (1989): Technology transfer: an assessment of the major institutional vehicles for diffusion of U.S. biotechnology. Technovation, 9, 463-478.

Senker, J.; Sharp, M. (1988): The Biotechnology directorate of the SERC. Report and Evaluation of its Achievements 1981-1987, SPRU working paper.

Sharp, M. (1985): The New Biotechnology, European Governments in Search of a Strategy, Sussex European Papers No. 15, SPRU, University of Sussex.

Sharp, M. (1989): Biotechnology in the UK. OTA Conference on Biotechnology in a global environment, Washington, SPRU mimeo.

Sharp, M. (1989): The management and coordination of biotechnology in the U.K. 1980-88, Phil. Trans. R. Soc. Lond., B 324, 509-523.

Sharp, M. (1993): The Protein Engineering Club. An Assessment of the Club as an Experiment in Promoting a Programme of Strategic Research.

Tylecote, A. (1995): Environment, Technology and Economic Growth: Challenge to Sustainable Development, Edward Elgar.

ANNEX

Annex I: Methodological Notes

Table A1: **Common Methodological Principles and Structure for Sectoral Reports of the Technological Audit Hungary (OECD)**

1. **Overview of sector**
    Number of firms, distribution
    Breakdown of sector
    Institutes
    R&D statistics

2. **Methodology description**
    Approach
    Sources of statistics
    Tools: questionnaire, interviews, inspections, groups/individuals
    Criteria for selecting firms and institutes
    Size of samples
    Selection of experts

3. **Technology segments**
    Definition

4. **Competitiveness**
    Definition
    Rating by markets (internal, external) and
    by output-type (final product, subcontract)
    Present and potential
    Technological dependency
    Environmental factors (constraints/opportunities)

5. **Restraining and stimulating factors**
    General and sector-specific
    Lack of qualified suppliers
    Underemployment
    Standards
    Business environment
    Management issues

6. **Company/institute action options**
    Strategies proposed by managers

7. **Government policy options**
    R&D development policy options
    Innovation policy measures
    Improvement of linkages
    Tax measures
    OMFB's role
    Human resources
    International relations

8. **Recommendations**

    Annexes to sectoral report:
    A. Summary of visits
    B. Contact details of key persons at all visited institutes and companies in the sector
    C. Coordinates of experts participating in the sectoral Audit

Table A2: **Methodology of Technology Audit**
(Draft by the OECD Secretariate February 1995)

WHY?

- Identify competitive strengths and weaknesses
- Bring out policy options at government and at firm level
- Facilitate foreign investment

WHO?

- Open internationally to ensure fair competition
- Involve appropriate government institutions from the start
- Involve firms, institutes and related institutions
- Involve national experts
- Involve technical and management expertise
- High quality of experts is the key factor

HOW?

- Core elements: interviews and site visits by several experts
- Operate in local language
- Use international standards in data collection
- Be explicit about criteria and definitions (sectors, segments, competitiveness)
- Selection of firms should include some poor

WHAT?

- Equal time for evaluation and data collection
- Submit draft for comments from firms
- Iterative process with government
- Discuss effect of selection and non-answers
- Name actors in recommendations
- Follow-up actions should be embodied in process

INFORMATION?1

- Open information should be shared
- Confidentiality should be decided by orginators
- Reports in two parts:
    - one open
    - one closed (e.g. for firms)

Annex II: Revised Hungarian Intellectual Property Rights

Pharmaceutical Inventions

As the protection of medicaments and products produced chemically is prohibited, at present only the manufacturing processes are patentable; the protection extends, however, to the products directly obtained by means of the process.

A strongly amended law on patent protection for chemical and pharmaceutical products (Law No. VII of 1994) was enacted on July 1, 1994. This law introduced in Hungary product protection for chemically produced products, medicines and foodstuffs. Transitional patent protection (called pipeline protection) can be obtained on the basis of a patent granted abroad.

Inventions based on the first, second or further medical indication of a known compound are patentable as processes. For the second and any further medical indication protection may be granted if for a person skilled in the art the newly recognised effect does not follow obviously from the medical effect already known.

The important provisions of the law mentioned are the following:

(1) The provision excluding from patent protection medicaments, chemical products and foods for human and animal consumption (with the exception of plant varieties and animal breeds) are waived, consequently these products become patentable. It is understood that microorganisms and other biological objects are also patentable per se.

(2) Pharmaceutical products claimed in a foreign patent granted within one year following the entry into force of the above law and having a priority date falling between January 1, 1987 and July 1, 1993 are eligible for transitional protection (pipeline protection).

Pharmaceutical products marketed in Hungary prior to July 1, 1994 are excluded from transitional protection. The transitional protection is not enforceable against a third party who manufactured the pharmaceutical product in Hungary prior to July 1, 1994, if the manufacturer provides proof of the manufacturing with a document drawn up prior to the above date.

The term of transitional protection expires at the same time as the original term of the foreign patent.

The application for transitional protection shall be filed not later than within one year from the entry into force of the above law, i.e. by July 1, 1995. No restoration of rights is admissible in case of non-compliance with this time limit.

(3) Pending applications for inventions relating to processes for producing chemical products (filed prior to the effective date of the law amending the Patent Law) can be amended before a final decision is taken by the Office, by claiming also product protection. Such an amendment can, however, be effective only with a modified priority date of July 1, 1994. A request for an amendment can be filed not later than one year from the effective date of the above law. No restoration of rights is admissible in case of non-compliance with this time limit.

With these amendments, Hungary harmonised patentability of pharmaceutical and chemical products with US and European Union laws.

The Protection of Plant Protective

In this field the following inventions are patentable in Hungary:

- process for producing the active ingredient (compound)
- plant protective compositions
- process for producing plant protective
- process for utilising plant protective.

Since the production of an active ingredient always involves chemical reactions, patents may be granted only to manufacturing processes. On the contrary, the production of plant protective compositions hardly ever involves chemical reactions, or rather chemical reactions which are essential in obtaining the product, therefore patent protection may be granted for plant protective compositions if the application satisfies all the other criteria of patentability.

When in the patent application protection is claimed for the composition itself, the process for producing the composition may be patented only if in addition to the usual operations, certain special operations are also applied, and known operations are carried out in a special way, for example the claimed solution may be qualified as an activity not expected from a person skilled in the given art. In the case of utilisation of plant protective, protection should be claimed for the composition itself, the process of utilisation may be patented only if in addition to the usual operations or in the execution of such operations, the process has at least one specific characteristic on the basis of which the claimed solution may be qualified as an activity not expected from a person skilled in the given art.

The amendments provided for by Law No. VII of 1994 came into force on July 1, 1994.

Biotechnological and Genetic Inventions

For biological processes it is important to provide a detailed description permitting a good reproducibility, to indicate the starting materials and to specify the steps of the process. The starting materials have to be available to the public. The accessibility of the biological object representing the starting material (e.g. strain, vector, cell-line, enzyme etc.) must be certified.

In the case of biological processes, particularly recombinant DNA techniques and production of monoclonal antibodies, chemical and biological objects per se cannot be protected, only the manufacturing processes thereof. As to "traditional" micro-biological processes, only the species given in the example can be covered by the scope of protection, the genus cannot be.

The amendments provided for by Law No. VII of 1994 entered into force on July 1, 1994.

Plant Varieties and Animal Breeds

Plant varieties and animal breeds are patentable under the special provisions laid down by Patent Law. These provisions fully conform to the International Convention for the Protection of New Varieties of Plants.

A plant variety is patentable if it is distinguishable, novel, homogeneous, stable and it has been given a variety denomination suitable for registration.

Under the patent granted for a plant variety the patentees exclusive right of exploitation extends to the production for purposes of commercial marketing, the offering for sale or the marketing of the propagating material of the plant variety and to the repeated utilisation of the plant variety for the production of another plant variety. The propagating material may be exported only by the authorisation of the patentee.

The patent protection has a duration of 18 years for vines and trees, and of 15 years for other plants. The patentee is obliged to maintain the plant variety during the period of the patent protection.

The patented plant variety may be put into public production only after a qualification by the State, as provided for in special legislation. The qualification is carried out by the Institute for Agricultural Qualification.
The National Office of Inventions registers the variety denomination together with the grant of a patent.

The patent granted for a plant variety has to be declared null and void with retroactive effect to its origin if the plant variety was not distinguishable or novel, or if its subject coincides with that of a patent having an earlier priority.

The provisions concerning the protection of plant varieties shall apply mutatis mutandis to animal breeds. The duration of patent protection shall be twenty years from the filing date.

Inventions Involving the Use of a Microorganism

Rule 21 (1) of the Decree on the implementation of the present patent law provides that "A patent application relating to an invention based on the use of a species of microorganism shall be accompanied by a certificate concerning the deposit of the said species the accessibility thereto shall be certified. Where the species is deposited after the date of filing the patent application, the date of deposit shall be regarded as the date of filing." According to Section 41 (1) of the patent law "The description shall make it possible for a person skilled in the art to carry out the invention on the basis of the description and drawings."

This means that a person skilled in the art can carry out inventions involving the use of microorganisms on the basis of the description if the microorganism has been marketed before the date of filing; it is disclosed in the description by reference to literature; it has been disclosed in an earlier patent description; or its production has already been described. If the microorganism is not available to the public, or it is new, the description must disclose the production of the microorganism in a manner sufficiently detailed for it to be carried out by a person skilled in the art. This can only be avoided if the microorganism has been deposited before the date of filing. If the deposit follows the date of filing and the description is insufficient, only the date of deposit can be accorded as date of filing.

The certificate of deposit indicates the place and date of deposit and refers to the depository institutions where the deposit was made. This certificate proves also that the microorganism furnished as sample by the given name and number is really a sample of the deposited microorganism, and that the deposited microorganisms can be preserved for the required time without changing their characteristics.

As known, according to the Budapest Treaty on the International Recognition of the Deposit of Microorganisms for the Purpose of Patent Procedure, concluded in 1977 and amended in 1980 any contracting State which allows or requires the deposit of microorganisms for the purposes of patent procedure recognises the deposit of a microorganism with any international depository authority. In Hungary the National Collection of Agricultural and Industrial Microorganisms at the University of Horticulture and Food Industry has been an international depository authority under Article 7 of the Budapest Treaty since June 1, 1986.

Annex III: List of Investigated Institutes and Firms

Firms

Agrar SH. Co.of Enying
    address: H-8155 Kiscsérpuszta
    phone: 36-22-367-876
    fax:36-22-367-801

Agricultural Sh.Co. of Szekszárd
    address: Szekszárd, Rákóczi út 132.
    phone: 36-74-316-438
    fax: 36--74-316-438

Agroferm
    address: H- 4181 Nádudvar P.O. Box 4.
    phone: 36-54-480-446, 36-54-480-465, 36-54-480-560
    Fax: 36-54-480-528
    contact person: dr. László Zahoránszky

Ceglédhús Meat Co.
    address: 2700 Cegléd, Dohány u. 30.
    phone: 36-53-311-442, 36-53-310-808
    fax: 36-53-311-562

Egis Pharmaceuticals Ltd. Biotechnical Pilot. Plant
    address: H-1092 Budapest, Hőgyes Endre u. 4.
    phone: 36-1-217-6852
    Fax: 36-1-217-6852
    contact person: Dr. Clara Ónody (head of Biotechnological Pilot Plant)

Győri "Keksz" Biscuits Ltd.

Agricultural Co-operative
    address: 9700 Szombathely, Vízöntő utca 9.
    phone: 36-94-313-576
    fax: 36-94-323-975
    contact person: Tibor Bindics, Enikő Bakos

Human Serum Production and Medicine Manufacturing SH. Co.

Hungarian Artificial Insemination Corporation
    address: H-2101 Gödöllő Nagyremete P.O. Box 74.
    phone: 36-28-410-337, 36-28-430-588
    fax: 36-28-410-337, 36-28-430-449
    contact person: Dr. Béla Zándoki

Jahn Ferenc Hospital
    address: 1204 Budapest, Köves út 2-4
    phone: 36-1-284-76-10
    fax: 36-1-284-76-57

Kaáli Institute
    address: Budapest, Istenhegyi út 54/a.
    phone: 36-202-28-02

Landhof Budapest Meat Ltd.
    address: 1097 Budapest Gubacsi út 6.
    phone: 36-1-215-39-40
    fax: 36-1-215-95-59

Medical Herb Research SH.Co.
    fax: 36-26-340-533

"Naszálytej" Milk SH.Co.
    address: 2600 Vác, Deákvári fasor 10.
    phone: 36-27-317-399
    fax: 36-27-314-997

Óbuda Nurseries Co-operative
Laboratory of Plant Pathology and Biotechnology
    address: H- 1039 Budapest, Királyok útja 226.
    phone: 36-1-3888-322
    fax: 36-1-250-4560
    contact person: Dr. Endre Tóth or Éva Kriston

Philaxia Vaccine Producing SH. Co.
    address: 1107 Budapest, Szállás utca 5.
    phone: 36-1-265-01-40
    fax: 36-1-261-41-21

Phylaxia-Sanofi Veterinary Biologicals Co. Ltd.
    address: H-1107 Budapest, Szállás u. 5.
    phone: 36-1-262-95-05, 36-1-261-2603
    fax: 36-1-260-38-89, 36-1-262-20-99
    contact person: Dr. Péter Sárközy

Recombi Vet Ltd.
    phone: 36-260-55-39
    fax: 260-3889

Saint László Hospital
    address: Budapest, Gyáli út 5-7.
    phone: 36-1-215-02-19
    fax: 36-1-215-0219

SKW Biotechnological Ltd.
    address: 1107 Budapest, Szállás utca 3.
    phone: 36-1-260-41-30
    fax: 36-1-262-4059

Institutes

Agricultural Research Centre of Martonvásár
    fax: 36-22-379-213

Biological Research Center of the Hungarian Academy of Sciences
    address: H-6701 Szeged, Temesvári krt. 62. P.O.Box 521, Hungary
    phone: 36-62-433-393, 36-62-432-232
    fax: 36-62-432-576
    e-mail: kulugy@everx.szbk.u-szeged.hu
    contact person: László Dallman, Ph.D.

Department of Applied Animal Genetics and Breeding
    phone: 36-1-322-26-60

DOTE Department of Biochemistry
    phone: 36-52-386-085

Enterprise for Research and Extension in Fruit Growing and Ornamentals
    fax: 36-1-226-80-37

Eötvös Lóránd University Department Plant Phisiology
    address: 1088 Budapest, Puskin utca 11-13.
    phone: 36-1-266-00-21
    fax: 36-1-266-00-21

Frucht Research Institute
    fax: 36-1-227-0979

Institute of Drug Research
    fax: 36-1-169-32-29

Institute of Enzimology
    address: Budapest, Karolina út 29.
    phone: 36-1-209-3535

József Attila University
Biotechnological Department
    address: 6726 Szeged, Temesvári krt. 48.
    fax: 36-62-454-352

Small Animal Research Institute
    fax: 36-28-330-184

University of Food Industry of Szeged
Biotechnological Department
    address: 6724 Szeged, Mars tér 7.
    phone: 36-62-454-000

University for Horticulture and Food Industry
Department of Chemistry and Biochemistry
    address: H-1518 Budapest, Villányi út 29-31. Hungary
    H-1518 Budapest, P.O. Box.53.
    phone: 36-1-185-33-22/228 or 239
    fax: 36-1-166-4272
    e-mail: jkosary@hoya.kee.hu
    contact person: D. Sci. Prof. Judit Kosáry

Research Institute for University of Veterinary Science
    adress: H-2225 Üllő, Dóra major
    contact person: Dr. Seregi János

Vegatable Crops Research Institute Co.
    address: H-1775 Pf.: 95. Hungary, Budapest
    phone: 36-1-226-78-32
    fax: 36-1-226-78-31
    contact person: Anikó Gémesné Juhász Dr.

Veterinary Medical Research Institute
Hungarian Academy of Sciences
    address: H-1581 Budapest, P.O. Box. 18.
    1143 Budapest, Hungária krt. 21.
    phone: 36-1-252-2455
    fax: 36-1-252-1069
    e-mail: Bnagy@novell.vmri.hu
    contact person: Dr. Béla Nagy

Bay Zoltán Foundation for Applied research
Institute for Biotechnology
    address: H-6726 Szeged, Derkovics fasor 2., Hungary
    phone: 36-62-432-251
    fax: 36-62-432-250
    e-mail: kalmanobay@szeged.hu
    contact person: Dr. Miklós Kálmán director

Semmelweis University of Medicine
    address: H-1085 Budapest, Üllõi út 26. III.em.
    phone: 36-1-117-29-79
    fax: 36-1-117-29-79
    e-mail: szokee@drog.SOTE.hu
    contact person: Dr. Éva Szõke

Experts

Dr. Maria PETZ-STIFTER, Hungarian Patent Office

Dr. Miklós NAGY, Undertaking Developing Centre, Budapest

Dr. Péter BIACS, Central Research Institute of Food Processing

Ferenc RUDÁN, OMFB

Sándor BOTTKA, OMFB

István BIHARI, OMFB Council

Dr. Béla RADOVICH, Ministry of Welfare

Dr. Erik BOGSCH, Chemicals Sh. Co. Richter Gedeon

TECHNOLOGY, INNOVATION and POLICY

Series of the Fraunhofer Institute
for Systems and Innovation Research (ISI)

Volume 1:
Kerstin Cuhls, Terutaka Kuwahara
Outlook for Japanese and German
Future Technology
1994. ISBN 3-7908-0800-8

Volume 2:
Guido Reger, Stefan Kuhlmann
European Technology Policy
in Germany
1995. ISBN 3-7908-0826-1

Volume 3:
Guido Reger, Ulrich Schmoch (Eds.)
Organisation of Science and Technology
at the Watershed
1996. ISBN 3-7908-0910-1

Volume 4:
Oliver Pfirrmann, Udo Wupperfeld and
Joshua Lerner
Venture Capital and New Technology
Based Firms
1997. ISBN 3-7908-0968-3

Volume 5:
Knut Koschatzky (Ed.)
Technology – Based Firms
in the Innovation Process
1997. ISBN 3-7908-1021-5

Volume 6:
Frieder Meyer-Krahmer (Ed.)
Innovation and Sustainable Development
1998. ISBN 3-7908-1038-X

Springer
and the
environment

At Springer we firmly believe that an
international science publisher has a
special obligation to the environment,
and our corporate policies consistently
reflect this conviction.
We also expect our business partners –
paper mills, printers, packaging
manufacturers, etc. – to commit
themselves to using materials and
production processes that do not harm
the environment. The paper in this
book is made from low- or no-chlorine
pulp and is acid free, in conformance
with international standards for paper
permanency.

Springer